# Fuzzy Recurrence Plots and Networks with Applications in Biomedicine

Tuan D. Pham

# Fuzzy Recurrence Plots and Networks with Applications in Biomedicine

Tuan D. Pham
The Center for Artificial Intelligence
Prince Mohammad Bin Fahd University
Al Khobar, Saudi Arabia

ISBN 978-3-030-37532-4 ISBN 978-3-030-37530-0 (eBook)
https://doi.org/10.1007/978-3-030-37530-0



This Springer imprint is published by the registered company Springer Nature Switzerland AG
The registered company address is: Gewerbestrasse 11, 6330 Cham, Switzerland

*To my wife, Thi Pham, for her love, her patience, and her tremendous support of my work beyond regular hours.*

# Foreword

In this monograph, three well-known methodological approaches are used, but for the first time in an originally combined manner. Network studies have their origin in the work by Leonhard Euler who solved the Königsberg bridge problem in 1736. Recurrence is an old concept mentioned in various religions and got its first scientific foundation by Henri Poincaré in 1885. Fuzzy sets and fuzzy logic were introduced by Lotfi A. Zadeh in 1965.

After clearly introducing these concepts in the first part, Tuan D. Pham presents his very recently developed combination of them in the form of fuzzy recurrence plots and fuzzy recurrence networks and illustrates their potential for paradigmatic model systems from nonlinear dynamics. The basic algorithms necessary for applications are also included which make them easy to apply. In the third part, he gives several examples of very successful applications of these techniques in biomedicine, ranging from cancer, via Parkinson's disease, to depression. They indeed show how promising these fuzzy recurrence methods will be for various further applications.

This very well-written monograph by Tuan D. Pham is the first systematic presentation of the very young and strongly evolving field of fuzzy recurrence. It provides constructive impacts for theorists in complex systems science as well as clearly practical hints for a broad range of specialists in biomedicine but also in various other areas, such as in engineering, physics, biology, geosciences, and socioeconomics, who analyze experiments with respect to observational data.

Berlin/Potsdam, Germany
November 2019

Jürgen Kurths

# Preface

This book aims to introduce recent developments in nonlinear data analysis known as fuzzy recurrence plots and fuzzy recurrence networks.

The concept of recurrence plots was introduced by Eckmann et al. in 1987 [1] as a visualization tool for measuring the time constancy of dynamical systems, which are characterized with large-scale topology and small-scale texture.

Recurrence plots were studied again by Casdagli ten years later in 1997 [2], who showed that recurrence plots can address some main objectives in nonstationary time series analysis, including (1) the characterization of the dynamics underlying a time series, (2) prediction of nonstationary time series, (3) identification of the change point at which the dynamics underlying a time series changes (change point detection), and (4) hypothesis testing for the null hypothesis that the time series is stationary.

Quantification of recurrence plots or recurrence quantification analysis (RQA) was introduced by Zbilut and Webber Jr. in 1992 [3]. In 2002, Marwan et al. [4] extended the RQA on vertical structures of recurrence plots and applied the RQA measures to the analysis of the logistic map and heart rate variability. Based on the theoretical framework of recurrence plots, recurrence networks were addressed by Donner et al. in 2010 [5], where the matrix of a recurrence plot is transformed into the adjacency matrix of a complex network.

The central rule for constructing a recurrence plot is the criterion for deciding if a pair of states of the dynamical system under study is similar. The notion of similarity lends itself to the utilization of fuzzy logic, which expresses similarity or closeness on a real continuous scale between 0 and 1 instead of 0 or 1 set by binary logic or the unit step function adopted in constructing a recurrence plot. In general, the construction of fuzzy recurrence plots avoids the rigidity for setting a similarity threshold, and fuzzy recurrence plots can deliver better visualization than recurrence plots. Furthermore, being represented as a grayscale image, a fuzzy recurrence plot allows the quantification or extraction of its features by many feature extraction methods developed in the rich literature of image processing. Fuzzy recurrence networks can be obtained from fuzzy recurrence plots and scalable to effectively handle long time series and big data.

Applications of fuzzy recurrence plots and their networks to several areas in biomedicine are presented in this book to demonstrate the potential of the new approach for the analysis of complex data in time series and images for machine learning and pattern recognition.

To provide the general readership with basic background for understanding of the mathematical formulations of fuzzy recurrence plots and networks, the book starts with a brief presentation of the phase space that lays a foundation for chaos and nonlinear time series analysis. Recurrence plots are then presented, including recurrence networks and recurrence quantification analysis. The concepts and operations on fuzzy logic together with the fuzzy *c*-means algorithms are described before the introduction of fuzzy recurrence plots and fuzzy recurrence networks. Some entropy methods developed for measuring the predictability or fluctuations of time series are also presented because they are discussed in the applications of fuzzy recurrence plots subsequently. These methods share a common criterion with recurrence plots for defining similarity between subseries of the original signal. The last chapter presents several applications in biomedicine, including the analysis of physiological signals, time series of mental-state dynamics, microscopy imaging, magnetic resonance imaging, and deep learning in artificial intelligence for analysis of short-time physiological time series.

Matlab codes for the fuzzy recurrence analysis are included in this book to facilitate the computer implementations of the developed algorithms and ideas.

Linköping, Sweden/Al Khobar, Saudi Arabia
December 2019

Tuan D. Pham

## References


1. Eckmann JP, Kamphorst SO, Ruelle D (1987) Recurrence plots of dynamical systems. Eur Lett 5:973–977
2. Casdagli MC (1997) Recurrence plots revisited. Phys D 108:12–44
3. Zbilut JP, Webber Jr. CL (1992) Embeddings and delays as derived quantification of recurrence plots. Phys Lett A 171:199–203
4. Marwan N et al (2002) Recurrence-plot-based measures of complexity and their application to heart-rate-variability data. Phys Rev E Stat Nonlin Soft Matter Phys 66:026702
5. Donner RV et al (2010) Recurrence networks—a novel paradigm for nonlinear time series analysis. New J Phys 12:033025

# Acknowledgements

First of all, I am grateful to Wayne Wheeler, who is the Senior Editor for Computer Science in Springer's London office, for his keen interest in learning the topic of my keynote talk on fuzzy recurrence analysis at the 10th International Conference on Bioinformatics and Biomedical Technology held in Amsterdam in 2018, and his kind suggestion to me to write this book.

Assistance and support from other members of Springer Nature are acknowledged herein: Simon Rees, Associate Editor, Computer Science; and Sriram Srinivas, Project Coordinator and Books Production.

The reviewers' feedback and comments on the book proposal were very helpful for shaping the structure and contents.

Jürgen Kurths to whom I am indebted for spending his valuable time in reading the draft of this book and writing the Foreword.

Tuan D. Pham

# Contents

# About the Author

**Tuan D. Pham** is Professor and Director of the Center for Artificial Intelligence at Prince Mohammad Bin Fahd University, Al Khobar, Saudi Arabia. The Center is co-funded by the Saudi Aramco. He was a Professor of Biomedical Engineering at Linköping University, Sweden, and held other previous positions as Professor and the Leader of the Aizu Research Cluster for Medical Engineering and Informatics, Head of the Medical Image Processing Laboratory at the University of Aizu, Japan; and the Bioinformatics Research Group Leader, the School of Engineering and Information Technology, the University of New South Wales, Australia. His teaching and research span across several disciplines of computer science and engineering. His current research areas include artificial intelligence, image processing, time series, nonlinear dynamics, novel methods in pattern recognition and machine learning applied to medicine, physiology, biology, and health. He has been actively serving as Section Editor and Associate Editor of several international journals, conference chair, keynote speaker, and invited speaker. Dr. Pham has been internationally recognized by his peers as a highly prolific scientist, publishing extensively as the lead author on various topics in books by well-known publishers, well-respected journals, and refereed conferences.

# Chapter 1
# Phase Space in Chaos and Nonlinear Dynamics

## 1.1 Phase Space

In chaos and nonlinear dynamics, the first basic consideration in time-series analysis is the reconstruction of a phase space or state space of the dynamical system (the terms "state" and "phase" will be used interchangeably). In other words, the analysis of a dynamical system relies on the concept of a phase space, which is a set of possible system states. The phase-space reconstruction in two or three dimensions allows the visualization of the dynamics. In the phase-space graph, a point plotted on the graph represents the state or phase of the system at a certain time. The time is expressed by the sequence of plotted points. The space on the graph is therefore called phase space or state space, in which the coordinates of the graph represent the variables needed to specify the phase or state of a dynamical system. A phase space with two axes (two phase/state variables) is called a two-dimensional phase space, and three axes (three phase/state variables), it is a three-dimensional phase space, and so on [1].

The state of a system can be described by its state variables as a function of time, that is, $s_1(t), s_2(t), \ldots, s_d(t)$, which form a vector in a $d$-dimensional space. For example, the Lorenz system [2] has $x$, $y$, and $z$ variables, which form a three-dimensional phase space. As the state changes with time, the vector in the phase space describes a trajectory representing the time evolution or the dynamics of the system. The shape of the trajectory tells about the underlying dynamics of the system, such as periodic or chaotic. The phase-space graph coupling with the trajectory of the dynamical system are called a phase-space portrait, phase portrait, or phase diagram [1].

Figure 1.1 shows the three components of the Lorenz system and its three-dimensional phase-space portrait.



T. D. Pham, *Fuzzy Recurrence Plots and Networks with Applications in Biomedicine*, https://doi.org/10.1007/978-3-030-37530-0_1

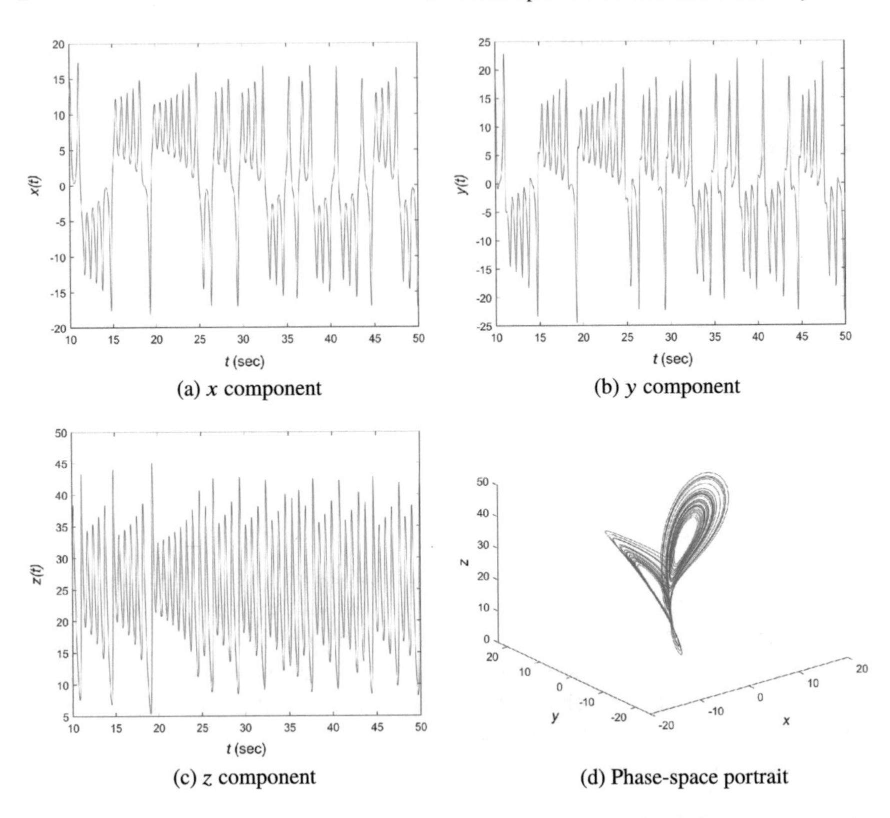


**Fig. 1.1** Three components of the Lorenz system and its three-dimensional phase-space portrait

## 1.2 Maps

The term "map" is often used in chaos theory. In mathematics, a map is a function. In chaos, a map is an iterative equation that specifies how a dynamical system evolves forward in time. It maps a value into another by specifying how a value of a variable takes a new value with an increased time step. For example, $s_{t+1} = 1 + s_t$ or $s_{t+1} = f(s_t)$. In general, given an initial value for $s_0$, the map iteratively outputs a sequence of new values for $s_1, s_2, s_3$, and so on. A one-dimensional map is a map that has only one physical feature, such as velocity.

## 1.3 Attractors

The abstract concepts of attractors play an important role for characterizing dynamical systems. There are several different definitions of an attractor [3]. One definition is that [4], in finite-dimensional systems, the coordinates of each point in the phase space refer to a collection of numerical values of the attributes that can be measured for a dynamical system. As time evolves, the combination of certain values of these attributes matches a region in a phase space. Such a region in a $d$-dimensional space is called an attractor. When a system is initialized with an unusual set of values of the attributes, which result in a point being far away from the attractor. As time travels across the phase space, the point quickly approaches the attractor. This behavior is therefore the reason it is named an attractor. The motion before the system reaches the attractor in the phase space is called transient. Furthermore, an attractor is called a strange attractor when it has a fractal dimension (a non-integer) [4].

## 1.4 Time Delay

In pseudo phase space, a time series is compared to later measurements within the same data. For example, a plot of $s_{t+1}$ versus $x_t$ shows how each observation $x_t$ compares to the next one $s_{t+1}$.

The amount of offset in the time series is called time delay. In other words, time delay, which is also called lag, is a constant time interval used as the basis for defining the subseries of the original one. For example, the subseries $s_{t+1}$ is based on the time delay of one, and $s_{t+2}$ is based on the time delay of two. The length of time-delayed series is $N - \tau$, where $N$ is the total number of observations in the original time series and $\tau$ is the time delay.

Table 1.1 shows an example of an original time series and its two subseries obtained from using time delays = 1 and 2.

**Table 1.1** Original time series and time-delayed series

| Time | Variable 1 | Variable 2 | Variable 3 |
|---|---|---|---|
| $t$ | $s_t$ | $s_{t+1}$ | $s_{t+2}$ |
| 1 | 10 | 20 | 30 |
| 2 | 20 | 30 | 40 |
| 3 | 30 | 40 | 50 |
| 4 | 40 | 50 | 60 |
| 5 | 50 | 60 | 70 |
| 6 | 60 | 70 | 80 |
| 7 | 70 | 80 | 90 |
| 8 | 80 | 90 | |
| 9 | 90 | | |
| Length | 9 | 8 | 7 |

## 1.5 Embedding Dimension

In graphical terms, the embedding dimension is the number of axes on a phase-space plot. From an analytical standpoint, the embedding dimension is the dimension of the phase space or number of state variables required to embed an object such as the trajectories of a dynamical system. From a prediction point of view, the embedding dimension is the number of successive points used to predict each next value in the time series [1]. The process of transforming a time series into an object in space is called embedding. Embedding a time series is a very important concept in chaos for reconstructing a phase space as it is a condition when the attractor in the original phase space is unfolded in the reconstructed phase space.

## 1.6 Phase-Space Reconstruction

For an analytical dynamical system, the phase space is known from the equations of motion. However, for an experimental dynamical system, its phase space is often unknown because of the lack of a mathematical description of the system. Phase-space reconstruction can reveal unobserved state variables because a single time series is the measurement of only one state variable. Therefore, attractor reconstruction methods have been developed for reconstructing the phase space of experimental systems.

Based on the Takens' embedding theorem [5], given appropriate values for a time delay and an embedding dimension, a single time series can be used to construct the observed state. This theorem works because the state variables of a dynamical system relate to the relationship that constructs the attractor [4]. The theorem states that if a time series comes from a dynamical system that is on an attractor, the trajectory constructed from the time series by embedding will have the same topological properties as the original one. This statement requires that if the original attractor has dimension $d$, then an embedding dimension of $m = 2d + 1$ will be sufficient for reconstructing the attractor. This reconstruction of the phase space is called time delay embedding.

The reconstructed subseries from the original time series $s(t)$ is expressed as

$$s^r(t_i) = [s(t_i), s(t_i + \tau), a(t_i + 2\tau), \ldots, s(t_i + (m-1)\tau)], \tag{1.1}$$

where $s^r(t_i)$ is the reconstructed state variable at discrete time $t_i$.

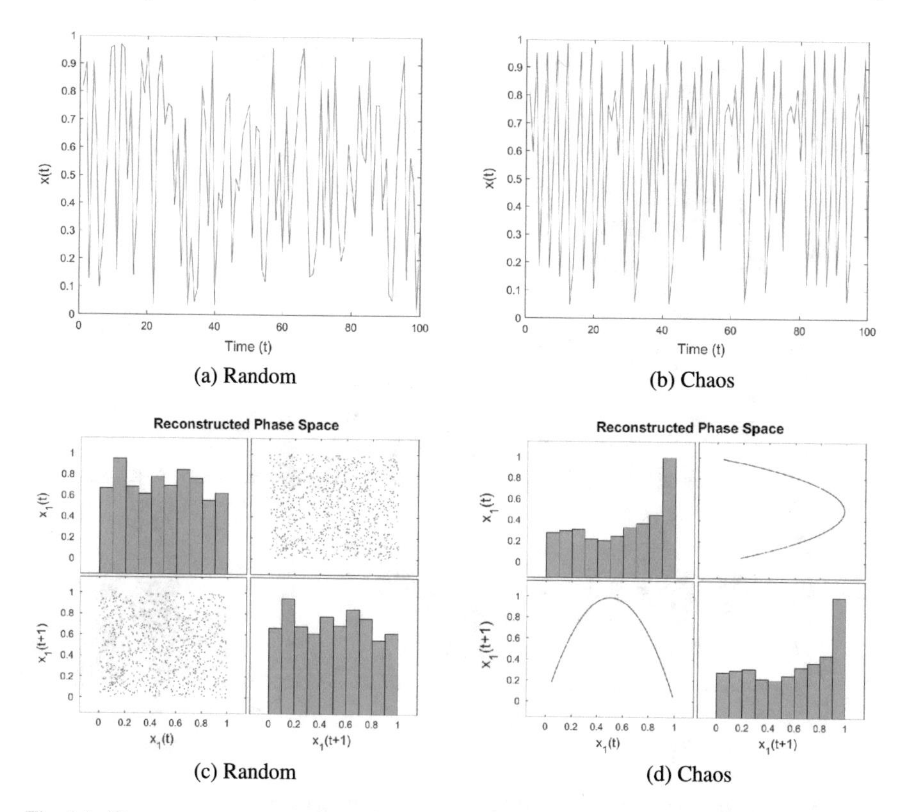


**Fig. 1.2** Phase-space reconstructions of random and chaotic (deterministic) time series

For example, for a time series $(s_1, s_2, s_3, \ldots, s_{10})$, the reconstructed attractor with $\tau = 3$, and $m = 2$ has subseries $(s_1, s_4)$, $(s_2, s_5)$, $(s_3, s_6)$, $(s_4, s_7)$, $(s_5, s_8)$, $(s_6, s_9)$, and $(s_7, s_{10})$ in a two-dimensional phase space.

Figure 1.2 shows the first 100 data points of two time series of 1000 data points generated with a uniformly distributed pseudorandom number function and a chaotic (deterministic) function $x(n+1) = 3.95x(n)[1 - x(n)]$ [4], where $x(1)$ of both time series = 0.8147. The diagonal plots from left to right in the figure are the corresponding histograms. The phase spaces of the two time series were reconstructed using time delay = 1, and embedding dimension = 2. Figure 1.2c, d are produced using the function "phaseSpaceReconstruction" of the Matlab R2019b Predictive Maintenance Toolbox. Although the fluctuations of the two time series look similar, the phase space of the random time series fills up the two-dimensional space, while the phase space of the chaotic time series does not fill up the two-dimensional space but forms a one-dimensional object called a strange attractor.

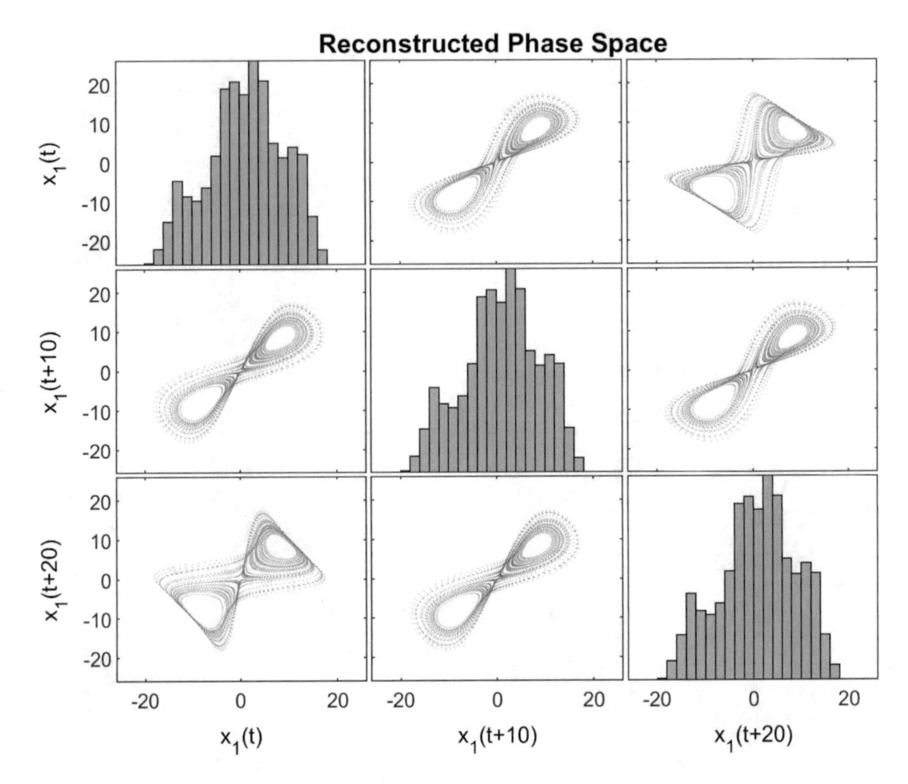


**Fig. 1.3** Phase-space reconstruction of the Lorenz system using the $x$-component

Several methods have been developed for estimating the best values for time delay and embedding dimension for the phase-space reconstruction [6]. For example, the time delay can be estimated using the method of average mutual information (AMI), where the best estimate for the time delay is set to be the first local minimum of AMI, and embedding dimension can be estimated using the false nearest neighbor (FNN) algorithm, where the estimated embedding dimension is the smallest value that satisfies the condition that the ratio of FNN points to the total number of points in the reconstructed phase space is less than the percentage of false neighbors.

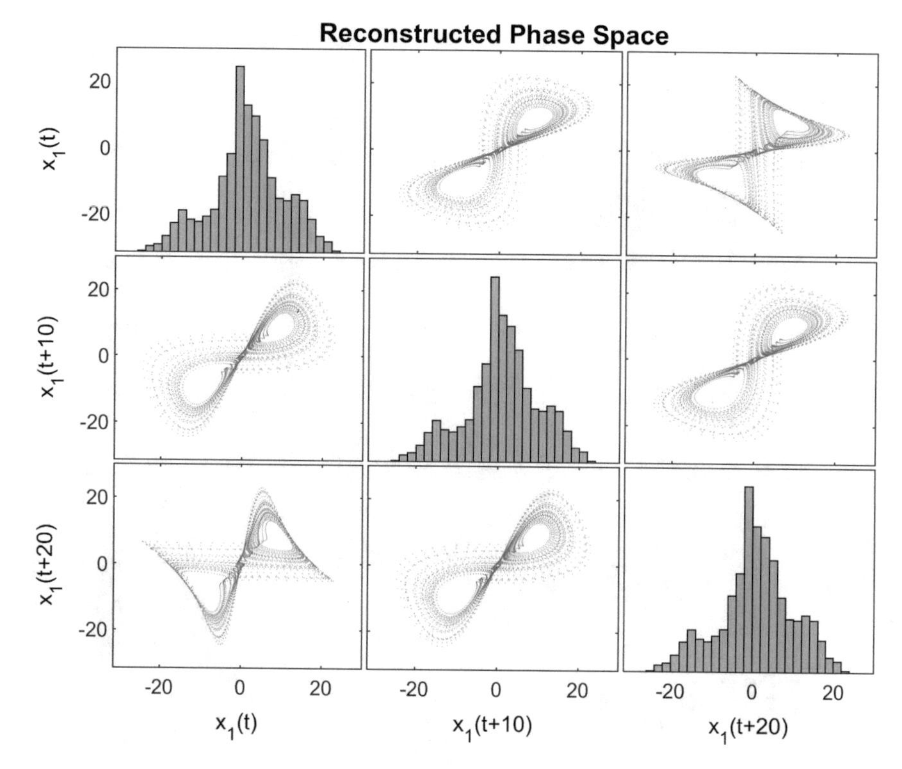


**Fig. 1.4** Phase-space reconstruction of the Lorenz system using the $y$-component

Figures 1.3, 1.4, and 1.5, which are produced using the function "phaseSpaceReconstruction" of the Matlab R2019b Predictive Maintenance Toolbox, show the phase-space reconstructions using each of the three components of the Lorenz system, where $\tau$ and $m$ are estimated as 10 and 3, using the AMI and FNN, respectively. As observed from the phase-space plot shown in Fig. 1.1, the topology of the attractor is recovered. The diagonal plots from left to right in Figs. 1.3, 1.4, and 1.5 represent the histograms of $x_1(t)$, $x_1(t+10)$, and $x_1(t+20)$ data, respectively.

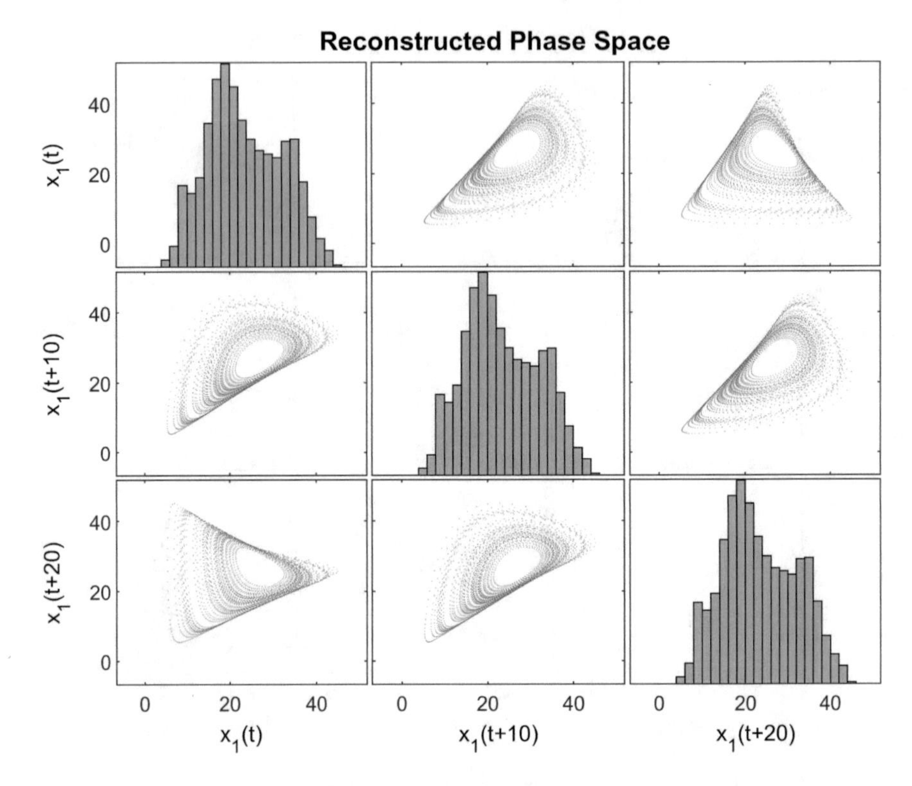


**Fig. 1.5** Phase-space reconstruction of the Lorenz system using the $z$-component

## References

1. Williams GP (1997) Chaos theory tamed. Joseph Henry Press, Washington, D.C
2. Lorenz EN (1963) Deterministic nonperiodic flow. J Atmos Sci 20:130–141
3. Milnor J (1985) On the concept of attractor. Comm Math Phys 99:177–195
4. Liebovitch LS (1998) Fractals and chaos simplified for the life sciences. Oxford University Press, New York
5. Takens F (1981) Detecting strange attractors in turbulence. Lect Notes Math 898:366–381
6. Kantz H, Schreiber T (2004) Nonlinear time series analysis. Cambridge University Press, Cambridge

# Chapter 2
# Recurrence Plots

## 2.1 Recurrence Plots

In chaos theory and nonlinear data analysis, the concept of recurrence plots was developed to show the times at which a phase-space trajectory visits the same area in the phase space [1]. Such revisits are called the recurrences that can reveal several interesting behaviors of a dynamical system.

Let $\mathbf{X} = \{\mathbf{x}_1, \mathbf{x}_2, \ldots, \mathbf{x}_N\}$ be a set of state vectors of the phase-space reconstruction of a dynamical system, in which $\mathbf{x}_i \in \Re^m$ is the $i$-a state of the dynamical system in an embedding dimension $m$. A recurrence plot (RP) is expressed an $N \times N$ matrix, in which an element of an RP, denoted as $\mathbf{R}(i, j)$, is represented with a black dot if the distance between $\mathbf{x}_i$ and $\mathbf{x}_j$ is considered to be closed to each other [1]. In other words, if $\mathbf{x}_i$ and $\mathbf{x}_j$ are similar, then there is an indication of recurrence. Two states $\mathbf{x}_j$ and $\mathbf{x}_i$ are said to be close or similar to each other if the distance between the two states is within the radius $\mathbf{r}_i$ whose center is $\mathbf{x}_i$ [1]. Because $\mathbf{r}_i$ and $\mathbf{r}_j$ may not be the same, this measure of similarity between two states of the phase-space reconstruction of a dynamical system can result in an asymmetrical RP.

To make an RP symmetrical, a threshold, denoted as $\epsilon$, is used to define the closeness or similarity of a state pair $(\mathbf{x}_i, \mathbf{x}_j)$ as follows [2]:

$$\mathbf{R}(i, j) = \theta(\epsilon - \|\mathbf{x}_i - \mathbf{x}_j\|), \tag{2.1}$$

where $\theta(\cdot)$ is the step function defined as

$$\theta(\epsilon - \|\mathbf{x}_i - \mathbf{x}_j\|) = \begin{cases} 1 : \epsilon - \|\mathbf{x}_i - \mathbf{x}_j\| \geq 0 \\ 0 : \epsilon - \|\mathbf{x}_i - \mathbf{x}_j\| < 0 \end{cases}. \tag{2.2}$$

A black dot is assigned to an element $(i, j)$ of an RP if $\mathbf{R}(i, j) = 1$, otherwise an RP element $(i, j)$ is represented with a white dot if $\mathbf{R}(i, j) = 0$. Thus, an RP can be visualized as a binary square image.

A historical review on the development of recurrence plots can be found in [3].



T. D. Pham, *Fuzzy Recurrence Plots and Networks with Applications in Biomedicine*, https://doi.org/10.1007/978-3-030-37530-0_2

To illustrate the construction of RPs, the stride times of three subjects, young, old, and Parkinson's disease (PD), obtained from the PhysioNet database [4] were used for constructing the RPs of the three nonlinear time series. The lengths of the time series for the young, old, and PD are 730, 815, and 222, respectively. Embedding dimension = 1 and time delay = 1 were used for the reconstruction of the phase space, and the similarity tolerance $\epsilon = 0.3$ used for constructing the RP. Figure 2.1 shows the time series of the three subjects. Figure 2.2 shows the RPs as an image and a sparsity pattern for the time series of each subject.

## 2.2 Cross and Joint Recurrence Plots

A cross recurrence plot (CRP) [5] can be visualized to express recurrences of simultaneous occurrences of phase-space trajectories of two dynamical systems, denoted as $\mathbf{x}_i$ and $\mathbf{y}_i$. A CRP is mathematically expressed as

$$\mathbf{C}RP(i,j) = \theta(\epsilon - \|\mathbf{x}_i - \mathbf{y}_j\|),$$
$$i = 1, \ldots, N, \; j = 1, \ldots, M, \tag{2.3}$$

where $\theta(\cdot)$ is the Heaviside step function , and $\epsilon$ is a threshold for computing the similarity or closeness between $\mathbf{x}_i$ and $\mathbf{y}_j$.

By definition, a CRP can be a non-square matrix if $N \neq M$. CRPs can be useful for the analysis of similarity of evolution or pattern matching between two different dynamical systems.

A joint recurrence plot (JRP) [6] graphically describes simultaneous occurrences of recurrences in two or more dynamical systems having the same number of states. In other words, a JRP is the Hadamard product of the RPs of multiple dynamical systems. A JRP of two dynamical systems is defined as

$$\mathbf{J}RP(i,j) = \theta(\epsilon_x - \|\mathbf{x}_i - \mathbf{x}_j\|) \cdot \theta(\epsilon_y - \|\mathbf{y}_i - \mathbf{y}_j\|),$$
$$i, j = 1, \ldots, N, \tag{2.4}$$

where $\epsilon_x$ and $\epsilon_y$ are two similarity thresholds prespecified for the two respective systems.

By definition, a JRP is a square matrix. JRPs can be useful for detecting generalized synchronization.

Figure 2.3 shows two components $x$ and $z$ of the Lorenz system [7], and their cross and joint recurrence plots presented as sparsity patterns. Parameters for constructing the plots are embedding dimension = 3, time delay = 2, $\epsilon = 0.2$, $\epsilon_x = 0.2$, and $\epsilon_z = 0.2$. Both CRP and JRP of the two Lorenz system components show a strong deterministic behavior indicated by the bands around the main diagonal lines.

**Fig. 2.1** Stride times recorded from young, old, and PD subjects

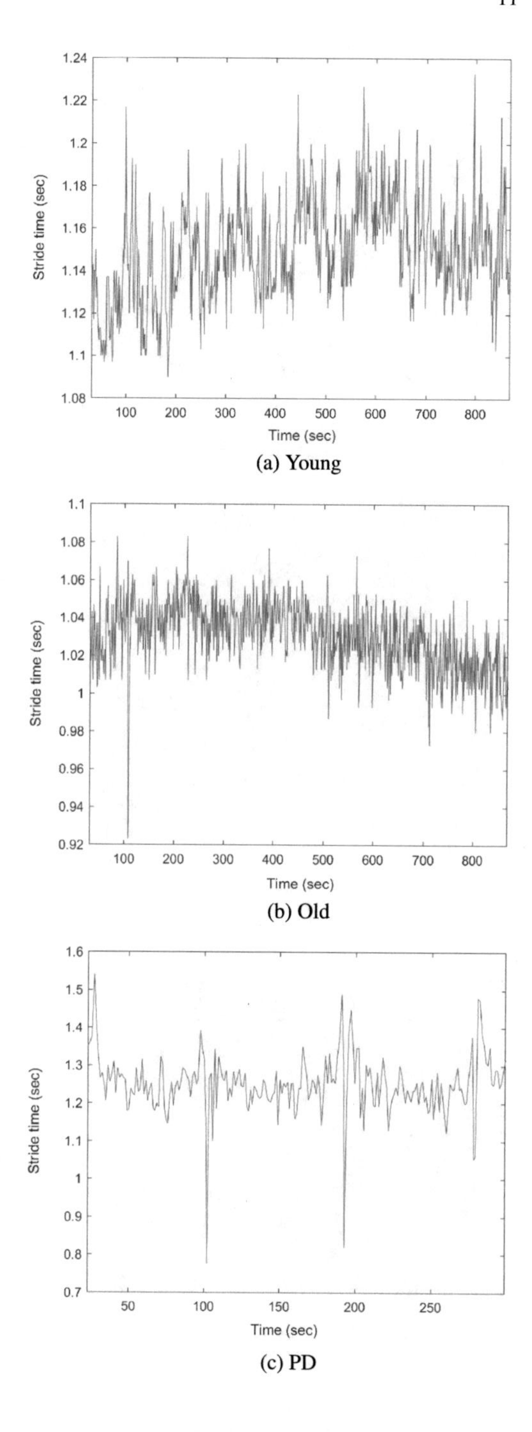


(a) Young

(b) Old

(c) PD

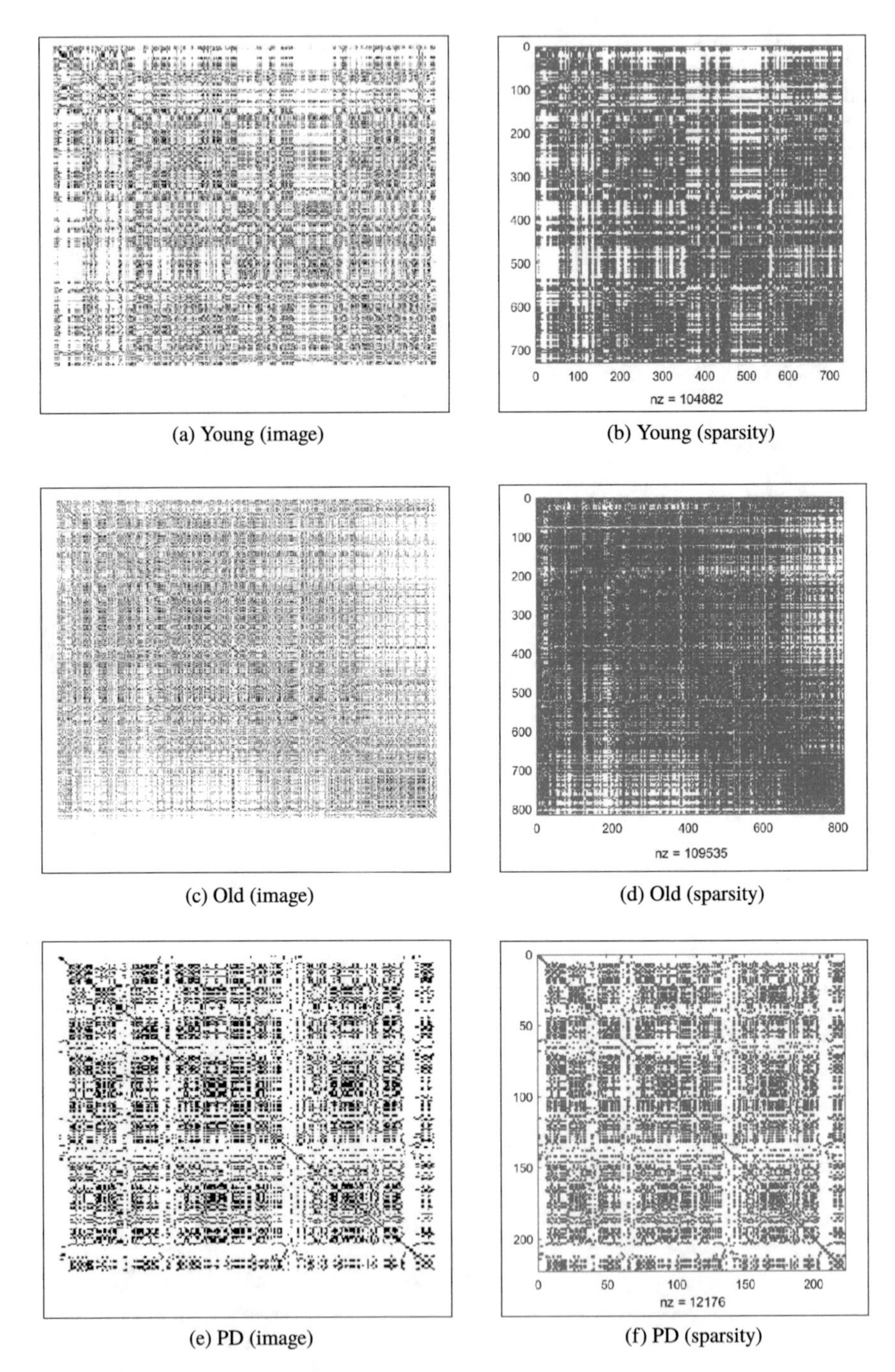


(a) Young (image) (b) Young (sparsity)

(c) Old (image) (d) Old (sparsity)

(e) PD (image) (f) PD (sparsity)

**Fig. 2.2** Recurrence plots of stride times recorded from young, old, and PD subjects visualized as grayscale images and sparsity patterns

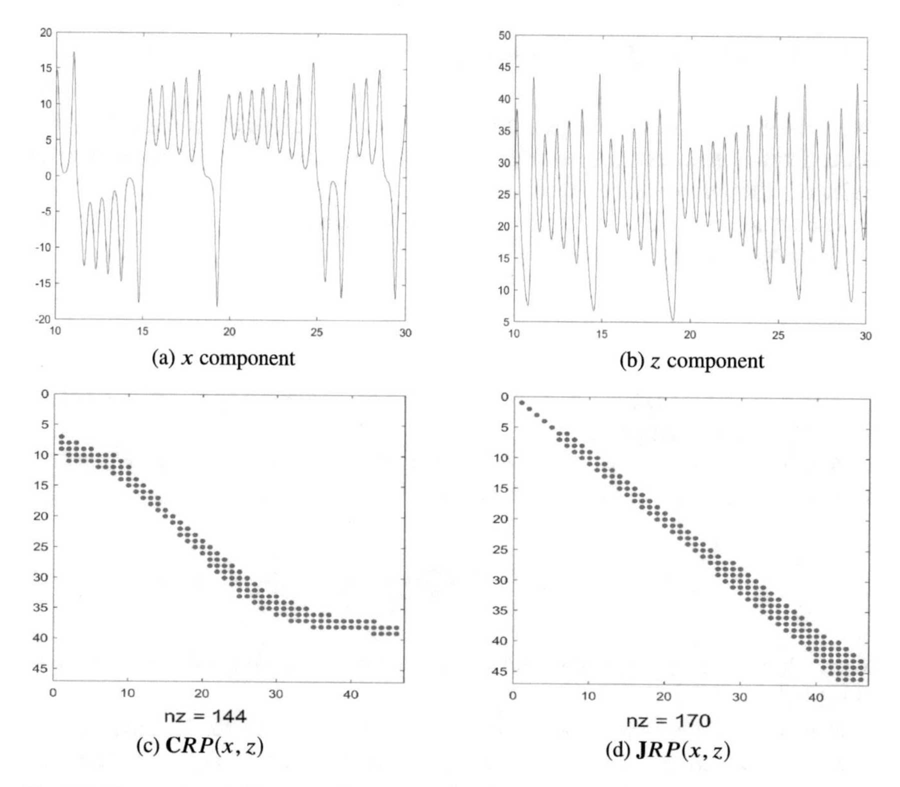

(a) $x$ component

(b) $z$ component

(c) $\mathbf{C}RP(x, z)$

(d) $\mathbf{J}RP(x, z)$

**Fig. 2.3** Time series of the Lorenz system and their cross and joint recurrence plots

## 2.3 Recurrence Networks

A recurrence network, whose elements are denoted as $\mathbf{A}_{ij}$, $i, j = 1, \ldots, N$, is a binary adjacency matrix defined in terms of the associated RP as follows [8]:

$$\mathbf{A}_{ij} = \mathbf{R}_{ij} - \delta_{ij}, \tag{2.5}$$

where $\delta_{ij}$ is the Kronecker delta. Alternatively, a recurrence network can be expressed as

$$\mathbf{A} = \mathbf{R} - \mathbf{I}, \tag{2.6}$$

where $\mathbf{I}$ is the identity matrix of size $N \times N$.

## 2.4 Recurrence Quantification Analysis

Recurrence quantification analysis (RQA) [2] is an approach that quantifies the number and duration of recurrences of a dynamical system presented by its phase-space trajectory (http://www.recurrence-plot.tk/rqa.php).

The recurrence rate (RR), which quantifies the density of recurrence points in an RP, is defined as

$$\mathrm{RR} = \frac{1}{N^2} \sum_{i,j=1}^{N} \mathbf{R}(i, j). \tag{2.7}$$

Equation (2.7) computes the recurrence rate as the probability of revisiting of the trajectory in the phase space.

Another measure of RQA is determinism (DET), which is the percentage of recurrence points which form diagonal lines:

$$\mathrm{DET} = \frac{\sum_{l=l_{\min}}^{N} l P(l)}{\sum_{l=1}^{N} l P(l)}, \tag{2.8}$$

where $P(l)$ is the frequency distribution (histogram) of length l of the diagonal lines of the RP.

Equation (2.8) is called determinism because it concerns with the predictability of a dynamical system. The RP of white noise contains almost single dots and very few diagonal lines, whereas the RP of a deterministic process consists of few single dots but many long diagonal lines.

The amount of recurrence points which form vertical lines, called laminarity (LAM), can be quantified in a similar way:

$$\mathrm{DET} = \frac{\sum_{v=v_{\min}}^{N} l P(v)}{\sum_{v=1}^{N} l P(v)}, \tag{2.9}$$

where $P(v)$ is the frequency distribution (histogram) of length v of the vertical lines of the RP.

The average diagonal line length (L) can be measured as

$$\mathrm{L} = \frac{\sum_{l=l_{\min}}^{N} l P(l)}{\sum_{l=l_{\min}}^{N} P(l)}. \tag{2.10}$$

Equation (2.10) measures the predictability time of the dynamical system.

In the same way, the trapping time (TT) measures the average length of the vertical lines. TT quantifies the laminarity time of the dynamical system. In other words, it indicates how long the system remains in a specific state. TT is defined as

$$\mathrm{TT} = \frac{\sum_{v=l_{\min}}^{N} l P(v)}{\sum_{v=l_{\min}}^{N} P(v)}. \tag{2.11}$$

Because the length of the diagonal lines of an RP is related to the time that the phase-space trajectory run parallel, the divergence behavior of the trajectories (DIV) is considered as the inverse of the longest diagonal line $L_{\max}$ (excluding the main diagonal line, which is also called the line of identity (LOI)). The DIV is stated to be related to the Kolmogorov–Sinai entropy of the system that is the sum of the positive Lyapunov exponents [9]. The DIV is defined as

$$\mathrm{DIV} = \frac{1}{L_{\max}} \tag{2.12}$$

The Shannon entropy of an RP can be used as a measure of recurrence and defined as

$$\mathrm{ENT} = -\sum_{l=l_{\min}}^{N} p(l) \ln(p(l)), \tag{2.13}$$

where

$$p(l) = \frac{P(l)}{\sum_{l=l_{\min}}^{N} P(l)}. \tag{2.14}$$

As another measure in RQA, the trend (TREND) provides information about the paling of the RP toward its edges, which is the effect of too low recurrence point densities in the edges of the RP. TREND is defined as

$$\mathrm{TREND} = \frac{\sum_{k=1}^{\tilde{N}} (k - \tilde{N}/2)(RR_k - < RR_k >)}{\sum_{k=1}^{\tilde{N}} (k - \tilde{N}/2)^2}, \tag{2.15}$$

where $< \cdot >$ is the average value, $\tilde{N} < N$, and

$$RR_k = \frac{1}{N-k} \sum_{j-i=k}^{N-k} \mathbf{R}(i, j). \tag{2.16}$$

## References


1. Eckmann JP, Kamphorst SO, Ruelle D (1987) Recurrence plots of dynamical systems. Europhys Lett 5:973–977
2. Marwan N, Romano MC, Thiel M, Kurths J (2007) Recurrence plots for the analysis of complex systems. Phys Rep 438:237–329
3. Marwan N (2008) A historical review of recurrence plots. Eur Phys J Spec Top 164:3–12
4. PhysioNet. https://www.physionet.org. Accessed 11 Feb 2018
5. Marwan N, Kurths J (2003) Nonlinear analysis of bivariate data with cross recurrence plots. Phys Lett A 302:299–307
6. Romano M, Thiel M, Kurths J, von Bloh W (2004) Multivariate recurrence plots. Phys Lett A 330:214–223
7. Lorenz EN (1963) Deterministic nonperiodic flow. J Atmos Sci 20:130–141
8. Marwan N, Kurths J (2015) Complex network based techniques to identify extreme events and (sudden) transitions in spatio-temporal systems. Chaos 25:097609
9. Quantification of recurrence plots (Recurrence Quantification Analysis). http://www.recurrence-plot.tk/rqa.php. Accessed 1 Sept 2019

# Chapter 3
# Fuzzy Logic

## 3.1 Fuzzy Sets

A fuzzy set is a collection of distinct objects whose membership grades in the set are expressed with real numbers [1]. Basic definitions of fuzzy sets are described as follows.

Let $X$ be a universe of discourse and $A$ a subset of $X$. A collection of generic elements of $X$ is $X = \{x\}$. A fuzzy set $A$ in $X$ is characterized by a fuzzy membership function $\mu_A(x)$ that maps each element in $X$ to a real number in the interval $[0, 1]$: $\mu_A(x) : X \to [0, 1]$. The real value of $\mu_A(x)$ is called the membership grade of $x$ in $A$. Thus, a greater value of the membership grade indicates a higher degree to which $x$ is a member of $A$.

By the above definition, a fuzzy set can be considered as an extension of a crisp set $B$ in $X$, whose crisp membership function is a mapping: $\mu_B(x) : X \to \{0, 1\}$, which indicates an element is either a member of the crisp set or not. In other words, a fuzzy set generalizes the concept of a crisp set by setting values of the membership function from the binary pair $\{0, 1\}$ to the real interval $[0, 1]$. Other properties of a fuzzy set can be defined as follows. A fuzzy set $A \in X$ is empty, denoted as $A = \emptyset$, if and only if (iff) its membership function is identically zero on $X$. Two fuzzy sets $A$ and $B$ are equal, denoted as $A = B$, iff $\mu_A(x) = \mu_B(x)$ or more simply $\mu_A = \mu_B$, $\forall x \in X$.

For example, let $X$ be the real number line $\Re$ and $A$ a fuzzy set of tall persons. The membership grades of some elements of $A$ representing height in meters can be subjectively given as $\mu_A(1.1) = 0$, $\mu_A(1.3) = 0.3$, $\mu_A(1.5) = 0.6$, $\mu_A(1.7) = 0.7$, and $\mu_A(2.0) = 1$.

Although the membership function of a fuzzy set is like a probability function when $X$ is a countable set, the meanings of the two functions are essentially different. A probability value expresses the frequency of occurrence in a crisp event, whereas a membership grade in fuzzy logic represents the degree of confidence in a fuzzy event. Thus, the concept of a fuzzy set is inherently nonstatistical.



T. D. Pham, *Fuzzy Recurrence Plots and Networks with Applications in Biomedicine*, https://doi.org/10.1007/978-3-030-37530-0_3

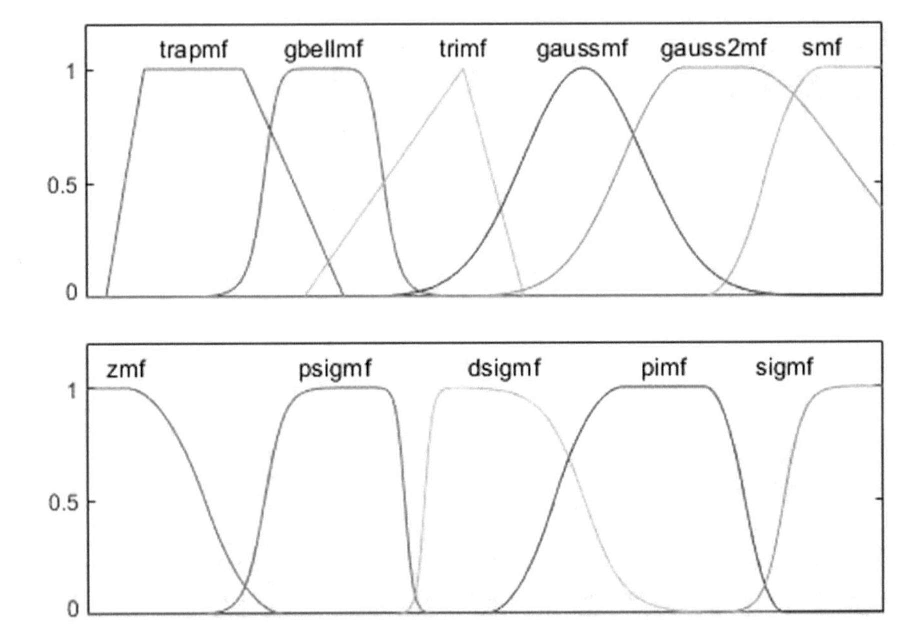


**Fig. 3.1** Typical fuzzy membership functions (trapmf = trapezoidal function, gbellmf = generalized bell curve function, trimf = triangular function, gaussmf = Gaussian function, smf = S-shaped curve function, zmf = Z-shaped curve function, psigmf = product of two sigmoid functions, dsigmf = function composed of the difference between two sigmoidal functions, pimf = pi-shaped curve function, and sigmf = sigmoid curve function)

Figure 3.1 shows 11 typical fuzzy membership functions of different shapes generated with certain parameters using the Matlab Fuzzy Logic Toolbox.

## 3.2 Operations on Fuzzy Sets

Some basic operations on fuzzy sets are given as follows [1, 2]:

- The complement of a fuzzy set A, denoted as $\bar{A}$, is defined as

$$\mu_{\bar{A}} = 1 - \mu_A. \tag{3.1}$$

- A fuzzy set $A$ is a subset of a fuzzy set $B$, denoted as $A \subset B$, iff $\mu_A \leq \mu_B$, which can be expressed as

$$A \subset B \Leftrightarrow \mu_A \leq \mu_B. \tag{3.2}$$

- The union of two fuzzy sets $A$ and $B$, denoted as $A \cup B$, results in a fuzzy set $C$ whose membership function is given as

$$\mu_C(x) = \max[\mu_A(x), \mu_B(x)], \forall x \in X, \tag{3.3}$$

or in an abbreviated form

$$\mu_C = \mu_A \vee \mu_B. \tag{3.4}$$

- The intersection of two fuzzy sets $A$ and $B$, denoted as $A \cap B$, results in a fuzzy set $C$ whose membership function is given as

$$\mu_C(x) = \min[\mu_A(x), \mu_B(x)], \forall x \in X, \tag{3.5}$$

or in an abbreviated form

$$\mu_C = \mu_A \wedge \mu_B. \tag{3.6}$$

- If a fuzzy set $A$ is a subset of a universe of discourse $X$, then an $\alpha$-level or $\alpha$-cut set of $A$ is a non-fuzzy set, denoted as $A_\alpha$, which comprises all elements of $X$ whose membership grades in $A$ are greater than or equal to $\alpha$. This is called a level set of a fuzzy set and is expressed as

$$A_\alpha = \{x : \mu_A(x) \geq \alpha\}. \tag{3.7}$$

## 3.3 Fuzzy Relations

A fuzzy relation $R$ from a set $X$ to a set $Y$ is defined as fuzzy subset of ordered pairs [3, 4]. If $X = \{x\}$ and $Y = \{y\}$ are collections of objects generically denoted by $x$ and $y$, then a fuzzy relation $R$ from $X$ to $Y$ is a fuzzy subset of the Cartesian product $X \times Y$. $R$ is characterized with a bivariate fuzzy membership function $\mu_R(x, y) \in [0, 1]$, which expresses the strength of relation of each pair $(x, y)$ in $R$.

The concept of a similarity relation in fuzzy logic is a generalized concept of an equivalence relation [3]. By definition, a similarity relation $S$ is a fuzzy relation that has the following three properties:

1. Reflexivity: $\mu_S(x, x) = 1, \forall x \in S$.
2. Symmetry: $\mu_S(x, y) = \mu_S(y, x), \forall x, y \in S$.
3. Transitivity: $\mu_S(x, z) = \vee_y[\mu_S(x, y) \wedge \mu_S(y, z)]$.

For example, if $O = X \times Y$, which is a relation from $X = \{x_1, x_2\}$ to $Y = \{y_1, y_2\}$, and $P = Y \times Z$, which is a relation from $Y = \{y_1, y_2\}$ to $Z = \{z_1, z_2\}$, with values of membership strength given in the matrix form as

$$O = \begin{bmatrix} 0.3 & 0.8 \\ 0.6 & 0.9 \end{bmatrix} \text{ and } P = \begin{bmatrix} 0.5 & 0.9 \\ 0.4 & 1 \end{bmatrix}.$$

Using the property of transitivity (max-min composition) of the fuzzy similarity relation defined above, the fuzzy relation $Q = X \times Z$ can be obtained as

$\mu_Q(x_1, z_1) = \max[\min(0.3, 0.5), \min(0.8, 0.4)] = 0.4,$
$\mu_Q(x_1, z_2) = \max[\min(0.3, 0.9), \min(0.8, 1)] = 0.8,$
$\mu_Q(x_2, z_1) = \max[\min(0.6, 0.5), \min(0.9, 0.4)] = 0.5,$ and
$\mu_Q(x_2, z_2) = \max[\min(0.6, 0.9), \min(0.9, 1)] = 0.9.$

Or in the matrix form,

$$Q = \begin{bmatrix} 0.4 & 0.8 \\ 0.5 & 0.9 \end{bmatrix}.$$

## 3.4 The Fuzzy *c*-Means Algorithm

The fuzzy $c$-means (FCM) [5] tries to minimize the following generalized least-squared error functional or objective function [6]:

$$F_w(\mathbf{U}, \mathbf{V}) = \sum_{k=1}^{N} \sum_{i=1}^{c} (\mu_{ik})^w \|\mathbf{x}_k - \mathbf{v}_i\|_A^2, \tag{3.8}$$

where
$\mathbf{X} = (\mathbf{x}_1, \ldots, \mathbf{x}_N) \subset \Re^n$ = the data,
$c$ = the number of clusters in $\mathbf{X}$, $2 \leq c < N$,
$\mu_{ik}$ = fuzzy membership grade of $\mathbf{x}_k$ in cluster $\mathbf{v}_i$, $\mu_{ik} \in [0, 1]$,
$w$ = fuzzy weighting exponent, $1 \leq w < \infty$,
$\mathbf{U} = c \times N$ fuzzy $c$-partition of $\mathbf{X}$, where the elements of $\mathbf{U}$ are the fuzzy membership grades of $N$ data points in $c$ clusters,
$\mathbf{V} = (\mathbf{v}_1, \ldots, \mathbf{v}_c\}$ = clusters of $\mathbf{X}$,
$\mathbf{v}_i = (v_{i1}, \ldots, v_{in})$ = center of cluster $i$,
$\|\cdot\|_A$ = induced $A$-norm on $\Re^n$, and
$A$ = positive-definite $n \times n$ weight matrix.

The squared distance between $\mathbf{x}_k$ and $\mathbf{v}_i$ expressed in Eq. (3.8) is computed in the $A$-norm as

$$d_{ik}^2 = \|\mathbf{x}_k - \mathbf{v}_i\|_A^2 = (\mathbf{x}_k - \mathbf{v}_i)^T A (\mathbf{x}_k - \mathbf{v}_i). \tag{3.9}$$

The fuzzy membership grades of the objective function are subject to

$$\sum_{i=1}^{c} \mu_{ik} = 1, k = 1, \ldots, N. \tag{3.10}$$

Let $\mathbf{C_X}$ = sample covariance matrix of dataset $\mathbf{X}$, $\{e_i\}$ = eigenvalues of $\mathbf{C_X}$, $\mathbf{D_X}$ = diagonal matrix whose each diagonal element $ii$ is the corresponding eigenvalue $e_i$, and $\mathbf{I}$ = identity matrix. Three popular norms for use with the objective function defined in Eq. (3.8) are

$$A = \begin{cases} \mathbf{I} = \text{Euclidean norm}, \\ \mathbf{D}_{\mathbf{X}}^{-1} = \text{diagonal norm}, \\ \mathbf{C}_{\mathbf{X}}^{-1} = \text{Mahalonobis norm}. \end{cases} \tag{3.11}$$

The minimization of the objective function of the FCM is numerically carried out by an iterative process of updating the fuzzy membership grades and cluster centers until the convergence or maximum number of iterations is reached. The fuzzy membership grades and cluster centers are iteratively updated as

$$\mu_{ik} = \frac{1}{\sum_{q=1}^{c} \left( \frac{\|\mathbf{x}_k - \mathbf{v}_i\|}{\|\mathbf{x}_k - \mathbf{v}_q\|} \right)^{2/(w-1)}}; \tag{3.12}$$

$$\mathbf{v}_i = \frac{\sum_{k=1}^{N} (\mu_{ik})^w \, \mathbf{x}_k}{\sum_{k=1}^{N} (\mu_{ik})^w}, \; i = 1, \ldots, c. \tag{3.13}$$

The iterative procedure of the FCM is outlined as follows:

1. Given $c$, $w$, step $t$, $t = 0, \ldots, T$, initialize matrix $\mathbf{U}^{(t=0)} = [\mu_{ik}]^{(t=0)}$.
2. Compute $\mathbf{v}_i^{(t)}$, $i = 1, \ldots, c$, using Eq. (3.13).
3. Update $\mathbf{U}^{(t+1)}$ using Eq. (3.12).
4. If $\|\mathbf{U}^{(t+1)} - |\mathbf{U}^{(t)}\| < \epsilon$ or $t = T$, stop. Otherwise, set $\mathbf{U}^{(t)} = \mathbf{U}^{(t+1)}$ and return to step 2.

## 3.5 Cluster Validity for the FCM

Cluster validity is a study that attempts to identify the best partition of a given dataset into several clusters. Cluster validity for fuzzy clustering is an area of its own research. Algorithmic developments have been being reported for measuring compactness within a cluster and separation between clusters so that an optimal number of clusters can be determined for unsupervised fuzzy clustering [7–10]. Some early developments of cluster validity measures are presented herein.

The original concept of cluster validity for fuzzy clustering appears to be based on the degree of separation between two fuzzy sets [2]. The first functional formulated for measuring cluster validity is the partition coefficient, denoted as $F(\mathbf{U}, c)$ and defined as [11]

$$F(\mathbf{U}, c) = \frac{1}{N} \sum_{i=1}^{c} \sum_{k=1}^{N} (\mu_{ik})^2, \tag{3.14}$$

where $c$ is the most plausible or optimal if its partition coefficient, $F(\mathbf{U}, c)$, is largest among a range of values for $c$.

The partition entropy, denoted as $H(\mathbf{U}, c)$, as a measure of cluster validity for the FCM is defined as [5]

$$H(\mathbf{U}, c) = -\frac{1}{N}\sum_{i=1}^{c}\sum_{k=1}^{N}\mu_{ik}\log(\mu_{ik}), \tag{3.15}$$

where $\log(\mu_{ik}) = 0$ for $\mu_{ik} = 0$, and $c$ is the most plausible if its partition entropy, $H(\mathbf{U}, c)$, is the smallest among a range of values for $c$.

As another measure of cluster validity, the proportion exponent, denoted as $P(\mathbf{U}, c)$, is defined as [12]

$$P(\mathbf{U}, c) = -\log\left(\prod_{k=1}^{N}\left[\sum_{j=1}^{[\mu_k^{-1}]}(-1)^{j+1}\binom{c}{j}(1 - j \vee_{i=1}^{c} u_{ik})^{c-1}\right]\right), \tag{3.16}$$

where

$$\mu_k = \max_{1\le i\le c}(u_{ik}), \tag{3.17}$$

and

$$[\mu_k^{-1}] = \text{ greatest interger } \le \frac{1}{\mu_k}. \tag{3.18}$$

The number of clusters $c$ is deemed to be optimal if its proportion component $P(\mathbf{U}, c)$ is largest among a given range of numbers of clusters.

It was pointed out that disadvantages of both $F(\mathbf{U}, c)$ and $H(\mathbf{U}, c)$ are their monotonicity and heuristic rationale. $F(\mathbf{U}, c)$ and $H(\mathbf{U}, c)$ are two extreme cases and can be proved to be equivalent. The proportion exponent $P(\mathbf{U}, c)$ tries to overcome the disadvantages of $F(\mathbf{U}, c)$ and $H(\mathbf{U}, c)$. However, the calculation of $P(\mathbf{U}, c)$ is rather stunning.

The partition index, denoted as $I(\mathbf{U}, c)$, is defined as [13]

$$I(\mathbf{U}, c) = \sum_{i=1}^{c}\frac{\sum_{k=1}^{N}(\mu_{ik})^w\|\mathbf{x}_k - \mathbf{v}_i\|^2}{N_i\sum_{q=1}^{c}\|\mathbf{v}_q - \mathbf{v}_i\|^2}, \tag{3.19}$$

where $N_i$ is the number of data points that have largest membership grades in cluster $\mathbf{v}_i$. The number of clusters $c$ is considered better than others if its partition index is the lowest.

The Xie-Beni index, denoted as $XB(\mathbf{U}, c)$, is defined as [14]

$$XB(\mathbf{U}, c) = \frac{\sum_{i=1}^{c}\sum_{k=1}^{N}(\mu_{ik})^w\|\mathbf{x}_k - \mathbf{v}_i\|^2}{N\min_{i,k}\|\mathbf{x}_k - \mathbf{v}_i\|^2}, \tag{3.20}$$

where the optimal number of clusters is the one whose index value is smallest.

## 3.6 Illustrations

One hundred uniformly distributed pseudorandom data points of two dimensions were generated. The function "fcm.mat" of the Fuzzy Logic Toolbox of Matlab was used to carry out the FCM analysis. Figure 3.2 shows the two datasets of one hundred uniformly distributed pseudorandom data points that were partitioned into two and three clusters by the FCM, respectively. Default parameters were adopted, where the fuzzy weighting exponent = 2, maximum number of iterations = 100, and norm = Euclidean. The Matlab code that produced the results shown in Fig. 3.2 is given below. By modifying the given Matlab code, the FCM analysis of another set of one hundred uniformly distributed pseudorandom data points with two clusters is also given in Fig. 3.2.

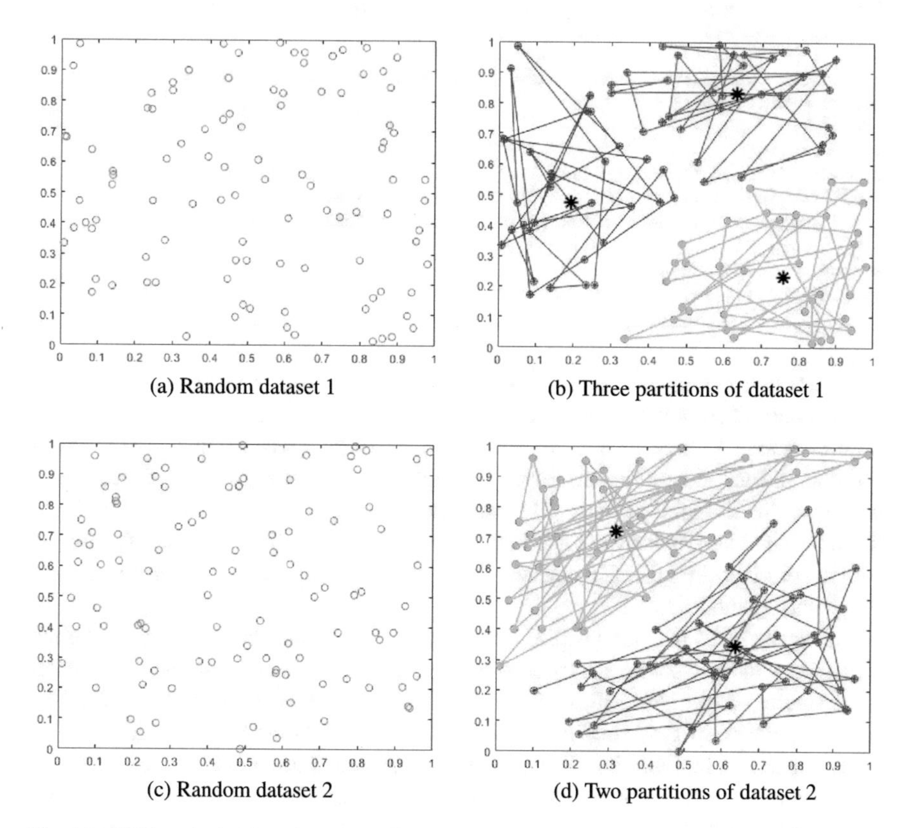


(a) Random dataset 1

(b) Three partitions of dataset 1

(c) Random dataset 2

(d) Two partitions of dataset 2

**Fig. 3.2** FCM analysis of random data, where "*" indicates a cluster center

### FCM analysis of random data

```
% Generate and plot data
data = rand(100,2);
plot(data(:,1), data(:,2),'o');
figure
plot(data(:,1), data(:,2),'o');
hold on;


% FCM with 3 clusters using default parameters
[center,U,obj_fcn] = fcm(data,3);
maxU = max(U);

% Find the data points with highest grade of membership in cluster 1
index1 = find(U(1,:) == maxU);

% Find the data points with highest grade of membership in cluster 2
index2 = find(U(2,:) == maxU);

% Find the data points with highest grade of membership in cluster 3
index3 = find(U(3,:) == maxU);

% Create lines
line(data(index1,1),data(index1,2),'marker','*','color','g');
line(data(index2,1),data(index2,2),'marker','*','color','r');
line(data(index3,1),data(index3,2),'marker','*','color','b');

% Plot the cluster centers
plot([center(:,1)],[center(:,2)],'*','color','k','MarkerSize',12,'LineWidth',2)
```

As another example, the Iris data [15] were used to illustrate the FCM analysis. The iris dataset contains random samples of flowers belonging to three species of iris flowers: setosa, versicolor, and virginica. For each of the species, the dataset contains 50 observations in centimeters for sepal length, sepal width, petal length, and petal width. In this illustration, only 10 observations for sepal length and sepal width of the setosa and versicolor were used for the sake of manual listing of the data, whose values are given in Table 3.1, and the fuzzy membership grades of each data point computed by the FCM.

Using the FCM provided by the Fuzzy Logic Toolbox of the Matlab with default parameters and two clusters, the coordinates of the two centers are obtained as: center 1 (versicolor) = (65.1461, 30.5773), and center 2 (setosa) = (49.0559, 31.3342). The fuzzy membership grades of the data points assigned to the two partitions by the FCM are shown in Table 3.2. The graphical results are shown in Fig. 3.3.

Using the dataset "fcmdata.dat" of the Matlab Fuzzy Logic Toolbox, which consists of 140 two-dimensional data points of two groups, Fig. 3.4 shows the two clusters obtained from the FCM analysis using the default parameters.

To illustrate the validity measures of the FCM analysis of the "fcmdata", the number of clusters from 2 to 10 were used to calculate the partition coefficient $F(\mathbf{U}, c)$, partition entropy $H(\mathbf{U}, c)$, partition index $I(\mathbf{U}, c)$, and the XB index $XB(\mathbf{U}, c)$. The results of the ten partitions and cluster validity measures for different numbers of

**Table 3.1** Ten observations for sepal length and sepal width of the setosa (denoted as 1) and versicolor (denoted as 2)

| Sepal length (cm) | Sepal width (cm) | Species |
|---|---|---|
| 51 | 35 | 1 |
| 49 | 30 | 1 |
| 47 | 32 | 1 |
| 46 | 31 | 1 |
| 50 | 36 | 1 |
| 54 | 39 | 1 |
| 46 | 34 | 1 |
| 50 | 34 | 1 |
| 44 | 29 | 1 |
| 49 | 31 | 1 |
| 70 | 32 | 2 |
| 64 | 32 | 2 |
| 69 | 31 | 2 |
| 55 | 23 | 2 |
| 65 | 28 | 2 |
| 57 | 28 | 2 |
| 63 | 33 | 2 |
| 49 | 24 | 2 |
| 66 | 29 | 2 |
| 52 | 27 | 2 |

**Table 3.2** Fuzzy membership grades of 20 observations for sepal length and sepal width of the setosa and versicolor assigned to two clusters by the FCM

| Data point | Center 1 | Center 2 |
|---|---|---|
| 1 | 0.0727 | 0.9273 |
| 2 | 0.0068 | 0.9932 |
| 3 | 0.0139 | 0.9861 |
| 4 | 0.0251 | 0.9749 |
| 5 | 0.0805 | 0.9195 |
| 6 | 0.2989 | 0.7011 |
| 7 | 0.0417 | 0.9583 |
| 8 | 0.0321 | 0.9679 |
| 9 | 0.0645 | 0.9355 |
| 10 | 0.0004 | 0.9996 |
| 11 | 0.9449 | 0.0551 |
| 12 | 0.9853 | 0.0147 |
| 13 | 0.9636 | 0.0364 |
| 14 | 0.3952 | 0.6048 |
| 15 | 0.9755 | 0.0245 |
| 16 | 0.5042 | 0.4958 |
| 17 | 0.9496 | 0.0504 |
| 18 | 0.1504 | 0.8496 |
| 19 | 0.9891 | 0.0109 |
| 20 | 0.1288 | 0.8712 |

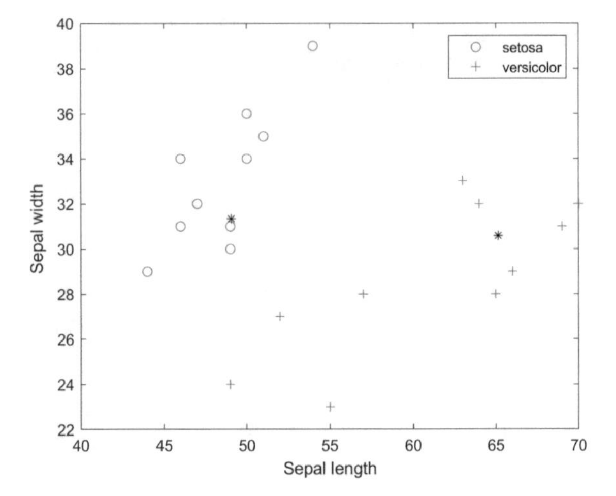


**Fig. 3.3** FCM analysis of Iris data subset, where "*" indicates a cluster center

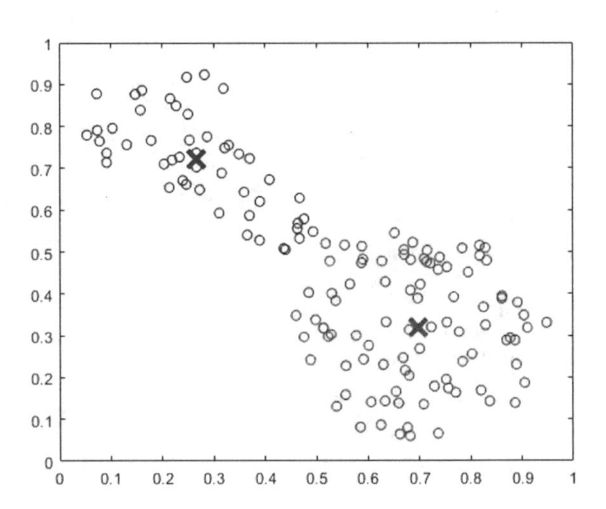


**Fig. 3.4** FCM analysis of the "fcmdata", where "x" indicates a cluster center

clusters are shown in Figs. 3.5 and 3.6, respectively. According to Fig. 3.6, the partition coefficient and partition entropy indicate the partition of the dataset into two clusters is optimal, while the partition index selects three clusters as the best partition, and the XB index suggests six clusters as the optimal number of clusters for the FCM partition among the range of ten clusters.

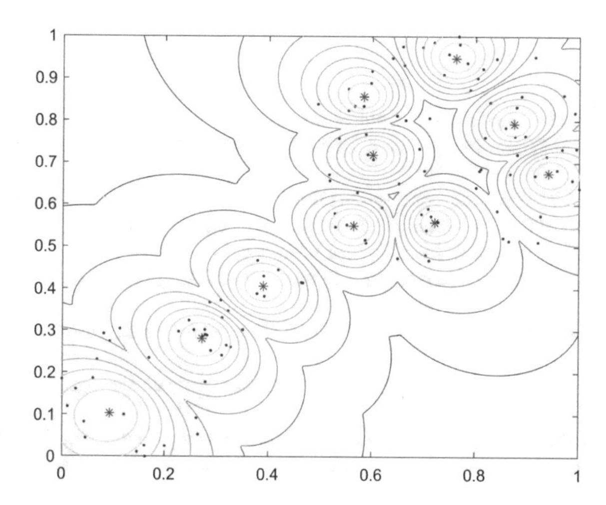


**Fig. 3.5** FCM analysis of the "fcmdata" with ten clusters, where "*" indicates a cluster center

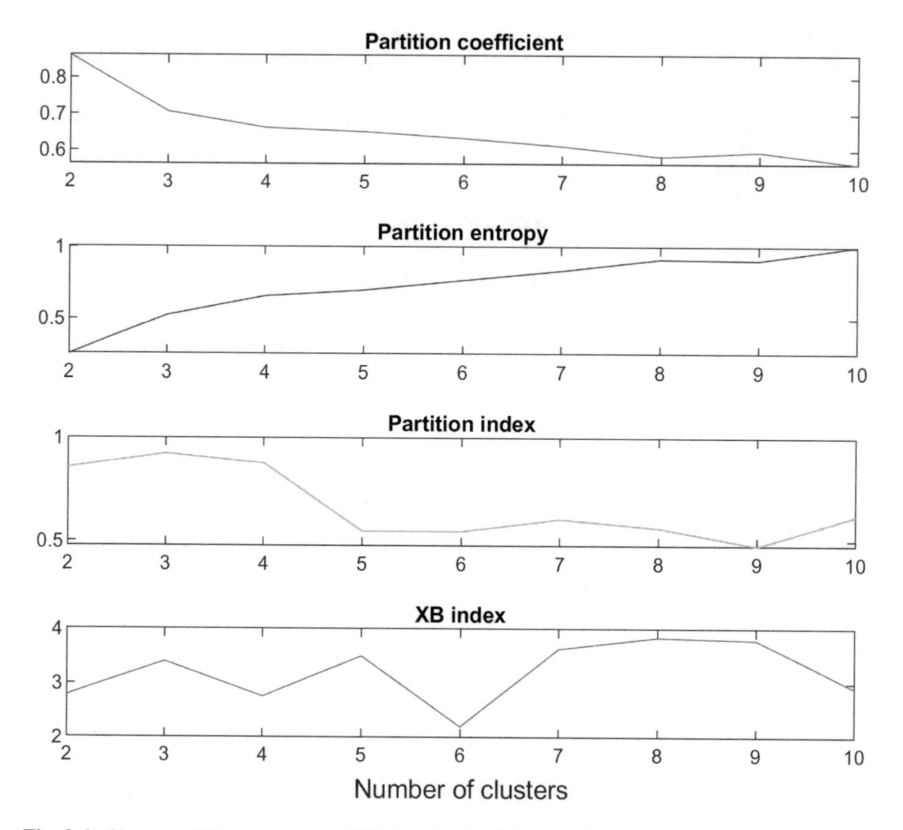


**Fig. 3.6** Cluster validity measures of FCM analysis of the "fcmdata" with ten cluster centers

## References

1. Zadeh LA (1965) Fuzzy sets. Inf Control 8:338–353
2. Zadeh LA (1975) The concept of a linguistic variable and its application to approximate reasoning-I. Inf Sci 8:199–249
3. Zadeh LA (1971) Similarity relations and fuzzy orderings. Inf Sci 3:177–200
4. Zadeh LA (1973) Outline of a new approach to the analysis of complex systems and decision processes. IEEE Trans Syst, Man, Cybern 3:28–44
5. Bezdek JC (1981) Pattern recognition with fuzzy objective function algorithms. Plenum Press, New York
6. Bezdek JC, Ehrlich R, Full W (1984) FCM: the fuzzy c-means clustering algorithm. Comput Geosci 10:191–203
7. Wu KL, Yang MS (2005) A cluster validity index for fuzzy clustering. Pattern Recognit Lett 26:1275–1291
8. Rezaee B (2010) A cluster validity index for fuzzy clustering. Fuzzy Sets Syst 161:3014–3025
9. Yang S et al (2018) A novel cluster validity index for fuzzy *c*-means algorithm. Soft Comput 22:1921–1931
10. Seghier ML (2018) Clustering of fMRI data: the elusive optimal number of clusters. Peer J 6:e5416
11. Bezdek JC (1974) Numerical taxonomy with fuzzy sets. J Math Biol 1:57–71
12. Windham MP (1981) Cluster validity for fuzzy clustering algorithms. Fuzzy Sets Syst 5:177–185
13. Bensaid AM et al (1996) Validity-guided (Re)Clustering with applications to image segmentation. IEEE Trans Fuzzy Syst 4:112–123
14. Xie XL, Beni GA (1991) Validity measure for fuzzy clustering. IEEE Trans Pattern Anal Mach Intell 3:841–846
15. Anderson E (1936) The species problem in Iris. Ann Mo Bot Gard 23:457–509

# Chapter 4
# Fuzzy Recurrence Plots

## 4.1 Fuzzy Recurrence Plots

### *4.1.1 Mathematical Formulation*

Let $\mathbf{X} = (\mathbf{x}_1, \ldots, \mathbf{x}_N)$ be a collection of the state vectors of the reconstructed phase space of a time series. Given $\mathbf{X}$, a predefined number of clusters $c$, and using the FCM algorithm [1, 2], a set of fuzzy clusters $\mathbf{V} = \{\mathbf{v}_1, \ldots, \mathbf{v}_c\}$ and the fuzzy membership grades expressing the degrees of the state vectors $\mathbf{x}_i \in \mathbf{X}$ belonging to cluster centers $\mathbf{v}_k \in \mathbf{V}$ can be determined to represent the partition of $\mathbf{X}$.

A fuzzy recurrence plot, denoted by $\tilde{\mathbf{R}}$, is defined as [3]

$$\tilde{\mathbf{R}}(i, j) = \mu(\mathbf{x}_i, \mathbf{x}_j), \; i, j = 1, \ldots, N, \tag{4.1}$$

where $\mu(\mathbf{x}_i, \mathbf{x}_j) \in [0, 1]$ is the fuzzy membership of similarity between $\mathbf{x}_i$ and $\mathbf{x}_j$, which can be inferred using the three properties of fuzzy relations [4] as follows:

1. Reflexivity:

$$\mu(\mathbf{x}_i, \mathbf{x}_i) = 1, \; i = 1, \ldots, N. \tag{4.2}$$

2. Symmetry:

$$\mu(\mathbf{x}_i, \mathbf{v}_k) = \mu(\mathbf{v}_k, \mathbf{x}_i), \; i = 1, \ldots, N, k = 1, \ldots, c. \tag{4.3}$$

3. Transitivity:

$$\mu(\mathbf{x}_i, \mathbf{x}_j) = \max[\min\{\mu(\mathbf{x}_i, \mathbf{v}_k), \mu(\mathbf{x}_j, \mathbf{v}_k)\}], \; k = 1, \ldots, c. \tag{4.4}$$

Based on the above formulation, a fuzzy recurrence plot (FRP), which is symmetrical and considered as the fuzzy relation matrix of the state vectors of the phase



T. D. Pham, *Fuzzy Recurrence Plots and Networks with Applications in Biomedicine*, https://doi.org/10.1007/978-3-030-37530-0_4

space. An FRP can be visualized as a grayscale image whose intensity values are represented with the complements of fuzzy relation matrix of the state-vector pairs $(1 - \tilde{\mathbf{R}})$ to be consistent with the display of a recurrence plot (RP). A darker pixel of an FRP, which is in the range [0, 1], has a stronger fuzzy membership grade and indicates that a state-vector pair is considered to be more similar. A black pixel of an FRP expresses that the two state vectors are the same (100% event of recurrence). The selection of the number of clusters for computing the FCM to construct an FRP is problem-dependent, which can be arbitrary or subject to the optimal criterion of one of fuzzy cluster validity methods introduced in Chap. 3.

### *4.1.2 Computer Code*

**Matlab Program for Constructing an FRP**

```
function FRP = frp(x,dim,tau,cluster,T)
%-----------------------------------------------------------------------
% Reference: Pham TD (2016) Fuzzy recurrence plots, EPL 116: 50008.
%-----------------------------------------------------------------------
% Input:
%       x: time series
%       dim: embedding dimension (default = 3)
%       tau: time delay (default = 1)
%       cluster: number of clusters (default = 2)
%       T: cuttoff fuzzy membership threshold to change grayscale to
%       black and white.
%       If T not given, FRP takes values in [0 1]. (default = NaN)
%
% Output:
%       FRP: Fuzzy recurrence plot
% Test:
%       x1 = randi([0 5],1,500); dim=3; tau=1; cluster=3; T=0.5;
%       FRP=frp(x1,dim,tau,cluster,T);
%       x2 = randi([0 255],1,500);
%       FRP=frp(x2);
%-----------------------------------------------------------------------

switch nargin
    case 1
        dim=3;
        tau=1;
        cluster=2;
        T=NaN;
    case 2
        tau=1;
        cluster=2;
        T=NaN;
    case 3
```

```
        cluster=2;
        T=NaN;
    case 4
        T=NaN;
end

[nRow,~] = size(x);

if (nRow==1)
    x = x';
    [nRow,~] = size(x);
end

M=nRow-(dim-1)*tau;
PS=zeros(M,dim);

for i=1:M
    for j=1:dim
        PS(i,j)=x(i+(j-1)*tau);
    end
end

[~, FR, ~] = fcm(PS, cluster); % use FCM from Matlab Toolbox

cRP=zeros(length(FR),length(FR),cluster);

for c=1:cluster
    for i=1:length(FR)
        for j=1:length(FR)
            if norm(PS(i,:) - PS(j,:)) == 0
                cRP(i,j,c)= 1; % reflexivity
            elseif FR(c,i)>=FR(c,j)
                cRP(i,j,c)=FR(c,j);
            else
                cRP(i,j,c)=FR(c,i);
            end
        end
    end
end


FRP=zeros(length(cRP),length(cRP));
for c=1:cluster
    for i=1:length(cRP)
        for j=1:length(cRP)
            if c==1
                FRP(i,j)=cRP(i,j,c);
            else
                if cRP(i,j,c)>=FRP(i,j)
                    FRP(i,j)=cRP(i,j,c);
                end
            end
        end
    end
end

if ~isnan(T)
```

```
    FRP(FRP>=T)=1;
    FRP(FRP<T)=0;
end

FRP = imcomplement(FRP); % Dark pixels to indicate recurrences

figure
imshow(FRP)

end
```

---

### 4.1.3 Illustrations

#### 4.1.3.1 The Lorenz System

Figure 4.1 shows 4000 time points of the $x$ (convection velocity), $y$ (temperature difference), and $z$ (temperature gradient) components of the Lorenz (chaotic) system constructed using standard parameters $\sigma = 10$, $\rho = 28$, and $\beta = 8/3$ [5]. These time series are used to illustrate advantages of FRP over RP with respect to visual effects, texture, and parameter specification for studying the recurrences of the dynamical system.

Using embedding dimension = 3 and time delay = 1, Figs. 4.2, 4.3, and 4.4 show the RPs and FRPs of the three components of Lorenz system with various values of the threshold as 10, 5, and 1% for the RPs, and various numbers of clusters as 2, 3, and 5 for the FRPs. The number of clusters in FRPs is inversely proportional to the threshold magnitude of RPs. This means setting a smaller number of clusters allows more states to be close to each other, making the plots denser with darker pixels, and being equivalent to the effect of setting a larger threshold of similarity for constructing RPs. The predefined FCM parameters $w$, maximum number of iterations, and tolerance for convergence are 2, 100, and 0.00001, respectively.

For all three components, changes in the visualization of the RPs are found to be more sensitive than those of the FRPs with respect to the variation of the associated parameters for constructing the respective plots. The RPs get fading with smaller thresholds of similarity (5 and 1% of the means) as being apparently noticed in the RPs of the $x$- and $y$-components, while the FRPs are well visible in displaying textural information about the degrees of recurrence of the state vectors in all three components. In general, it can be appreciated that the FRPs are more informative than the binary RPs in terms of visualization and texture of the characteristics of the dynamical system.

**Fig. 4.1** Time series of three components of the Lorenz system

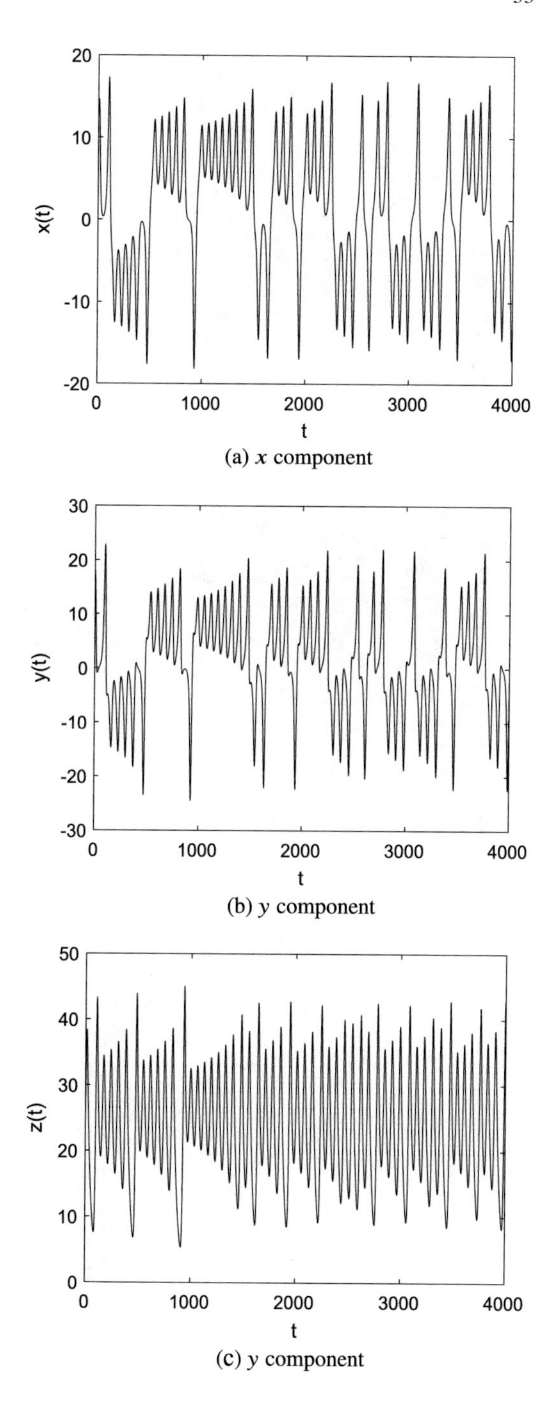


(a) $x$ component



(b) $y$ component



(c) $y$ component

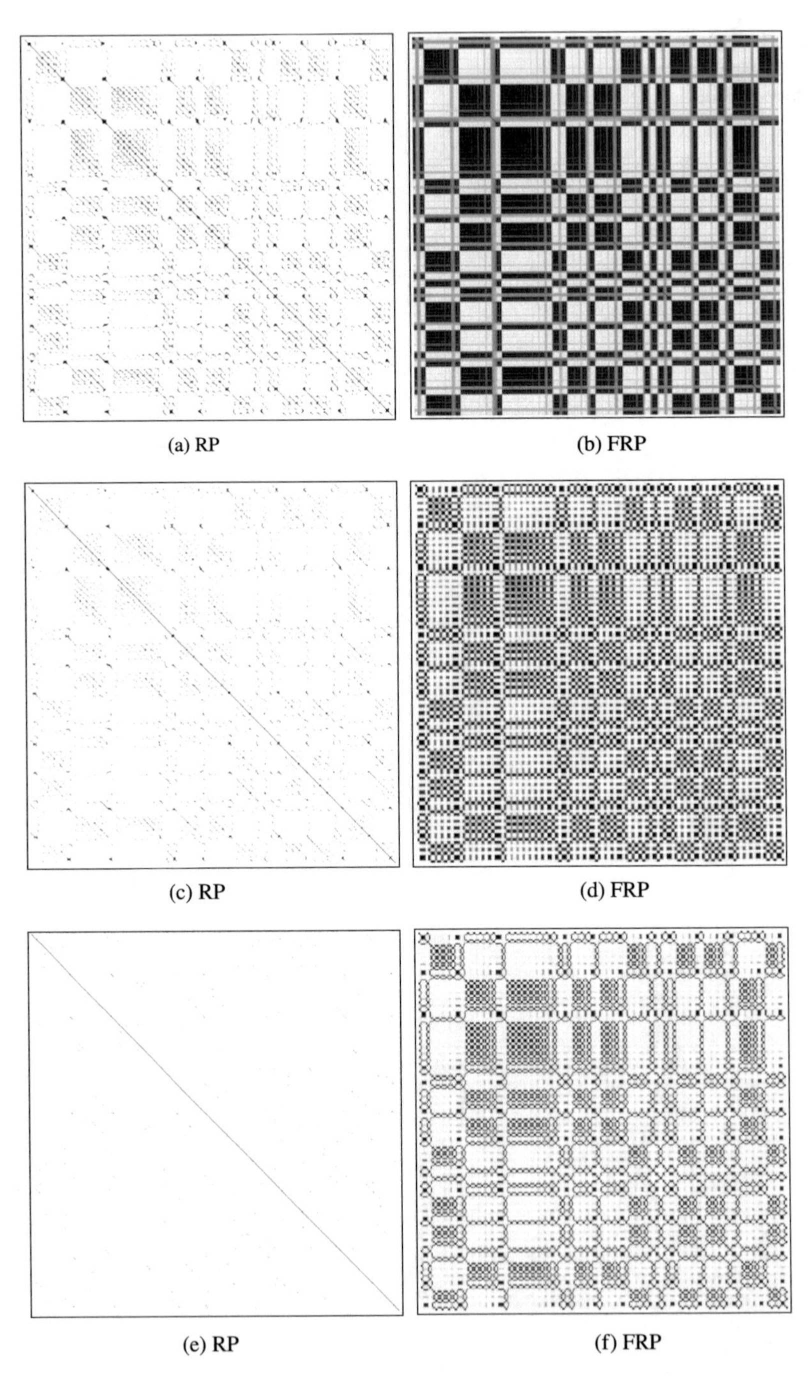


**Fig. 4.2** Recurrences of state vectors of the phase-space reconstruction of the $x$-component of the Lorenz system, with embedding dimension = 3, and time delay = 1: **a**, **c**, and **e** are recurrence plots with similarity threshold = 10%, 5%, and 1% of the mean, respectively; and **b**, **d**, and **f** are fuzzy recurrence plots with the number of clusters = 2, 3, and 5, respectively

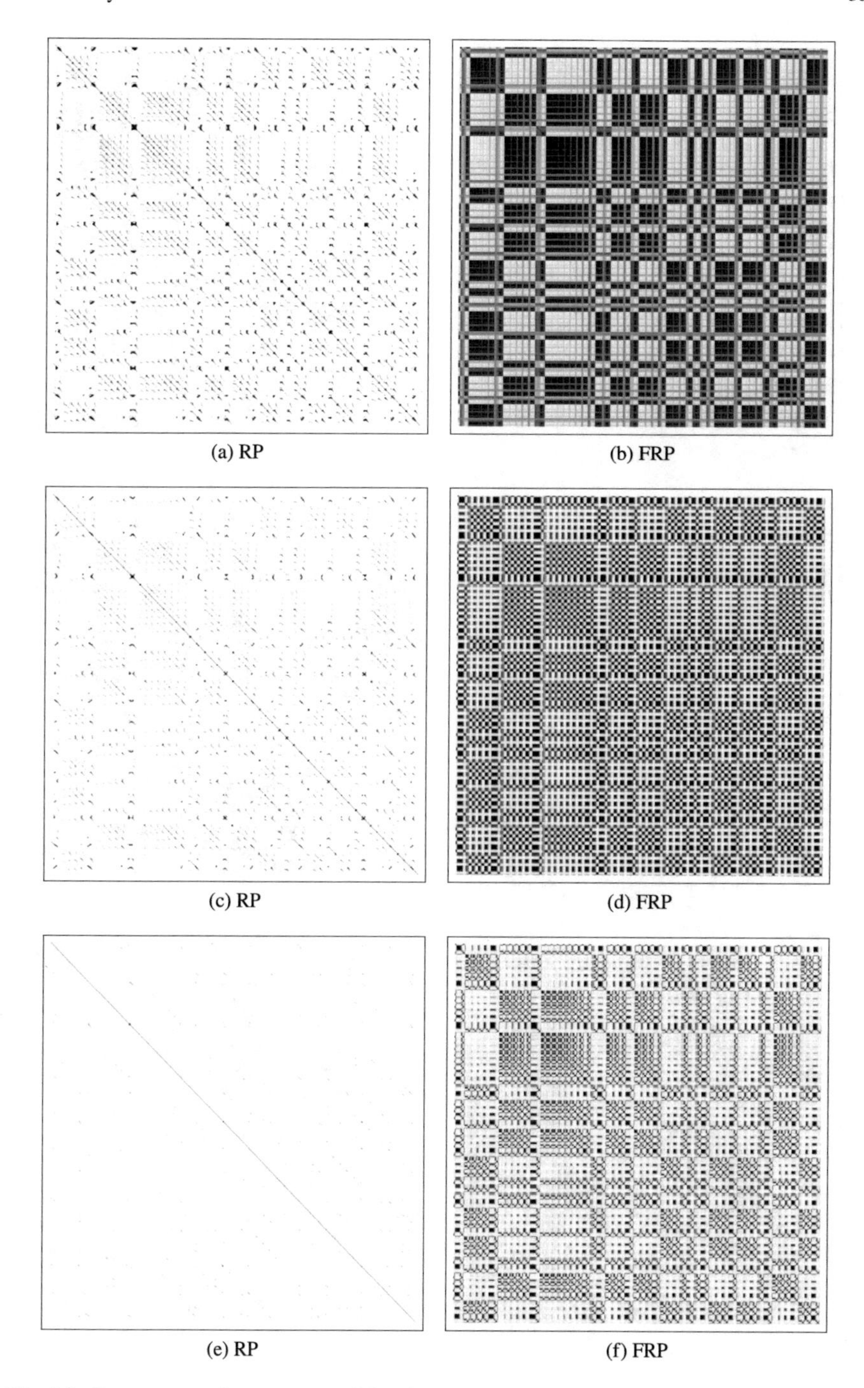


**Fig. 4.3** Recurrences of state vectors of the phase-space reconstruction of the $y$-component of the Lorenz system, with embedding dimension = 3, and time delay = 1: **a**, **c**, and **e** are recurrence plots with similarity threshold = 10%, 5%, and 1% of the mean, respectively; and **b**, **d**, and **f** are fuzzy recurrence plots with the number of clusters = 2, 3, and 5, respectively

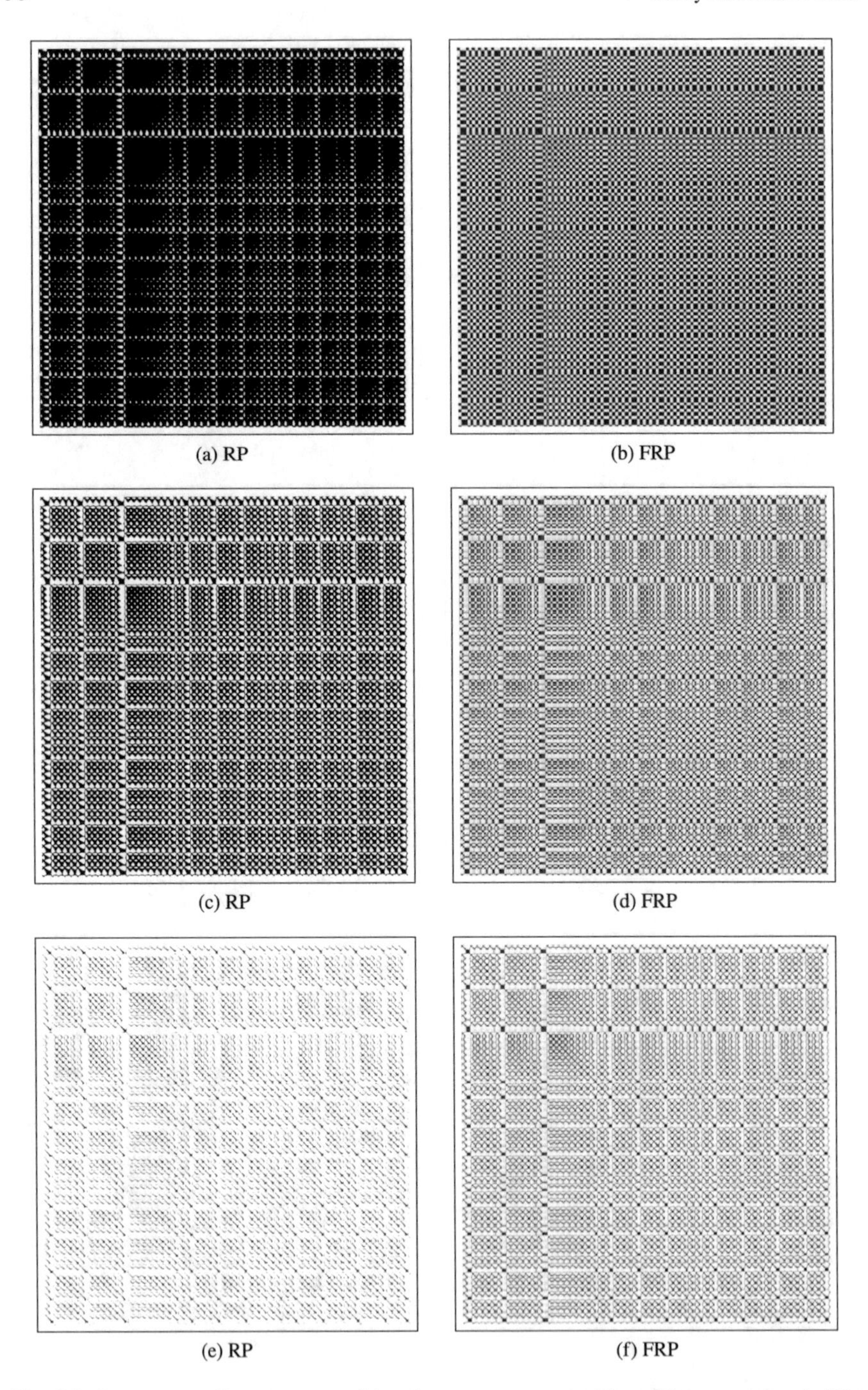


**Fig. 4.4** Recurrences of state vectors of the phase-space reconstruction of the $z$-component of the Lorenz system, with embedding dimension = 3, and time delay = 1: **a**, **c**, and **e** are recurrence plots with similarity threshold = 10%, 5%, and 1% of the mean, respectively; and **b**, **d**, and **f** are fuzzy recurrence plots with the number of clusters = 2, 3, and 5, respectively

#### 4.1.3.2 The Mackey–Glass Equation

Another chaotic time series is generated from the Mackey–Glass equation [6], which is a nonlinear time delay differential equation and defined as

$$\frac{dx}{dt} = \alpha \frac{x(t-\psi)}{1+[x(t-\psi)]^n} - \gamma x(t), \alpha, \gamma, n > 0, \tag{4.5}$$

where $\alpha$, $\gamma$, and $n$ are real numbers and selected as 0.2, 0.1, and 10, respectively.

When $x(0) = 1.2$, $\psi = 17$, and assuming $x(t) = 0$ for $t < 0$, Equation (4.5) produces a non-periodic and non-convergent time series that is very sensitive to initial conditions.

Figure 4.5 shows the plot of a Mackey–Glass time series whose length of 4000 time points is used in this illustration. To construct the phase space of the time series, the time delay and embedding dimensions were set to be 1 and 3, respectively. The predefined FCM parameters $w$, maximum number of iterations, and tolerance for convergence are 2, 100, and 0.00001, respectively.

Being like the cases of the Lorenz system, the visual displays of the recurrences of the dynamical system obtained from the FRPs are much clearer and less parameter-sensitive than those obtained from the RPs. The FRPs show much better textural contents than the RPs. Such texture can be useful for further development in the recurrence quantification (Fig. 4.6).

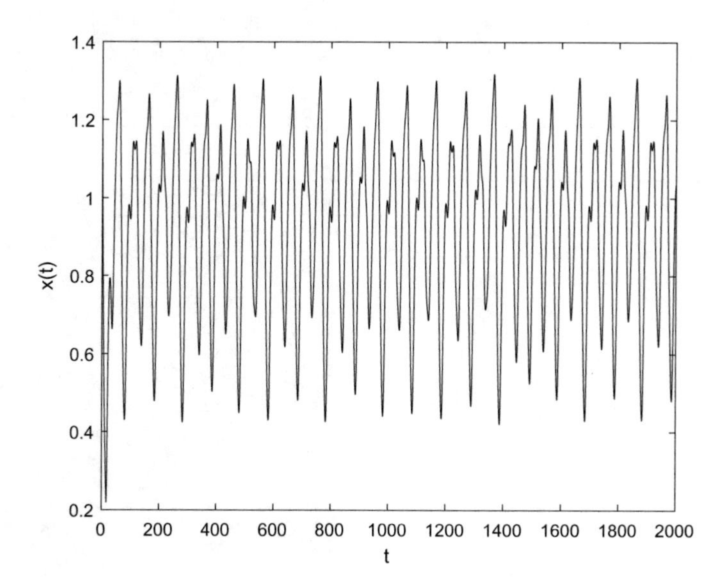


**Fig. 4.5** Mackey–Glass time series

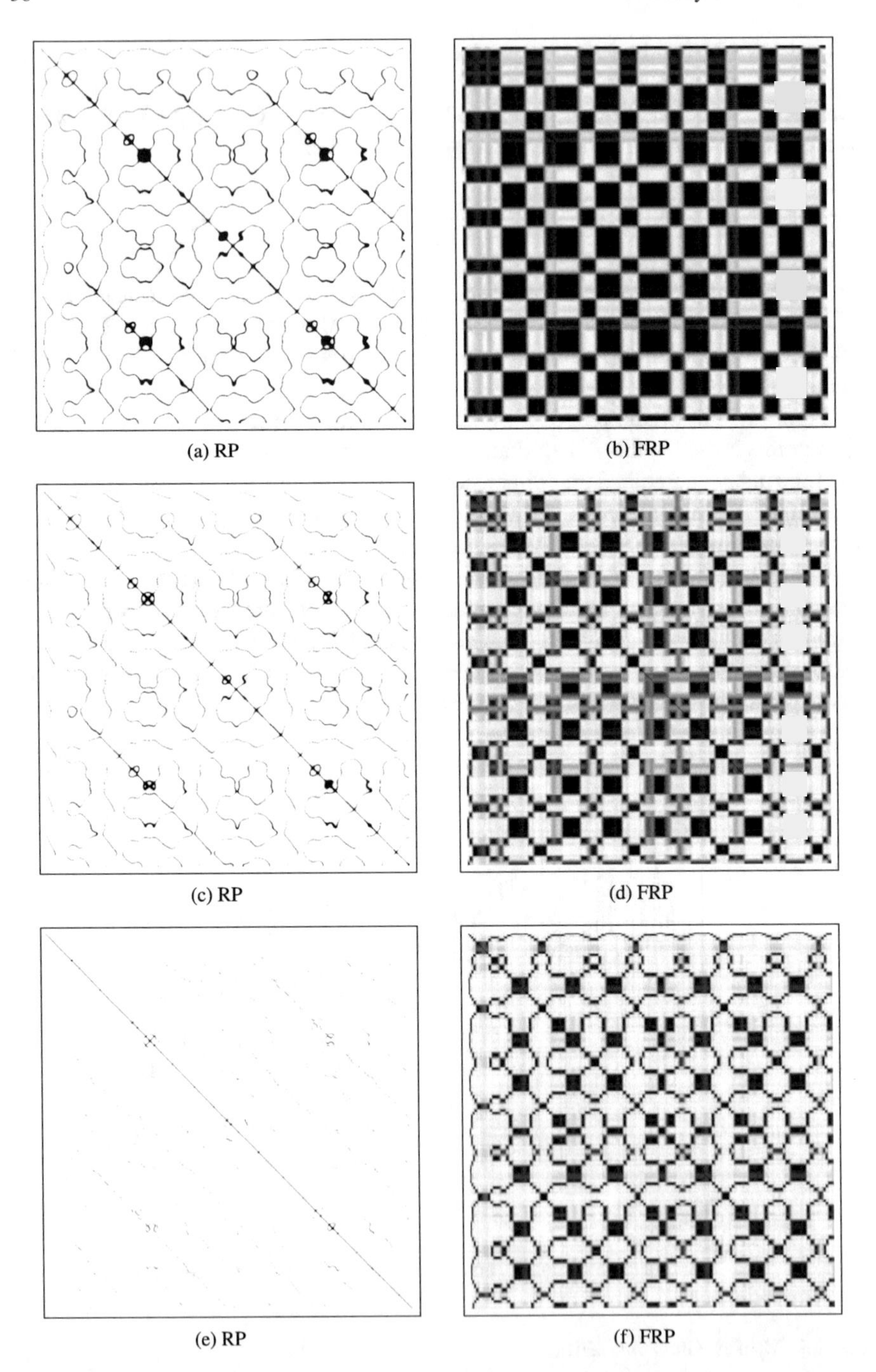


**Fig. 4.6** Recurrences of state vectors of the phase-space reconstruction of a Mackey–Glass time series, with embedding dimension = 3, and time delay = 1: **a**, **c**, and **e** are recurrence plots with similarity threshold = 10%, 5%, and 1% of the mean, respectively; and **b**, **d**, and **f** are fuzzy recurrence plots with the number of clusters = 2, 3, and 5, respectively

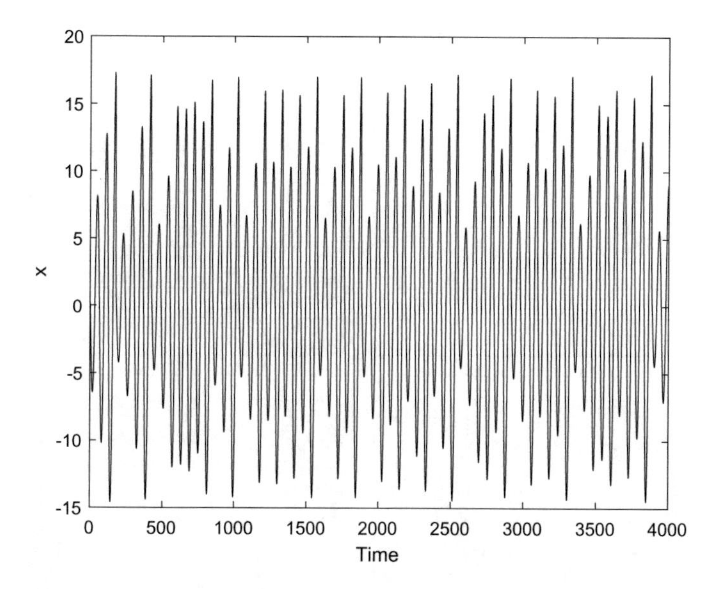


**Fig. 4.7** Time series of the Rossler attractor

#### 4.1.3.3 The Rossler System

A Rossler (chaotic) time series is used to illustrate the visual effects between the RPs and FRPs. The time series as shown in Fig. 4.7 was generated by the Runge–Kutta integration of the Rossler system of equations [7] expressed as

$$\frac{dx}{dt} = -z - y, \tag{4.6}$$

$$\frac{dy}{dt} = x + a \times y, \tag{4.7}$$

$$\frac{dz}{dt} = b + z \times (x - c). \tag{4.8}$$

The time series was generated using values for $a = 0.15$, $b = 0.20$, and $c = 10.0$. By observing the recurrence plots shown in Fig. 4.8, the FRPs provide much better visualization of the recurrences of the dynamical system than the RPs in terms of both structural patterns and texture that reveal an oscillating systems with diagonal-oriented, periodic recurrent arrangements of lines and dots of different gray levels.

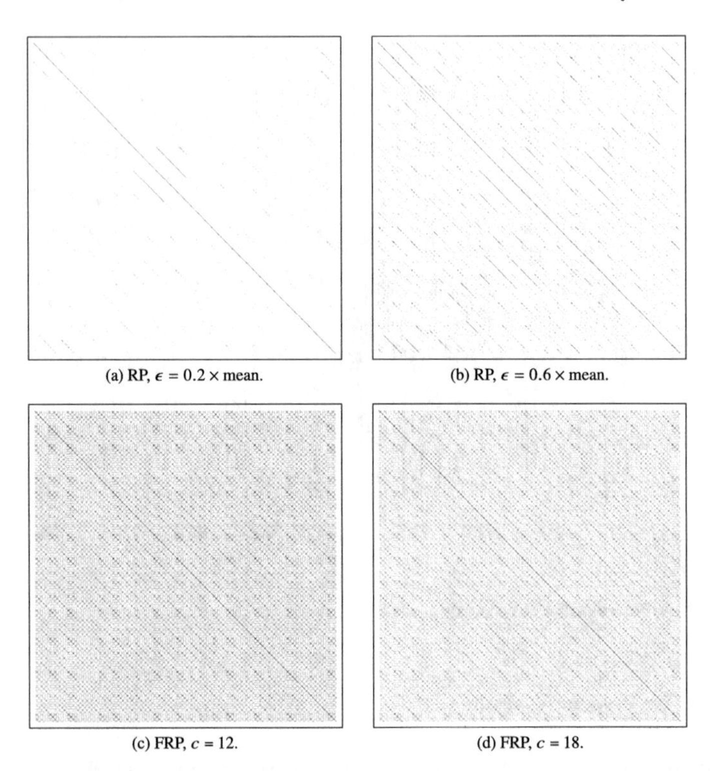

(a) RP, $\epsilon = 0.2 \times$ mean.

(b) RP, $\epsilon = 0.6 \times$ mean.

(c) FRP, $c = 12$.

(d) FRP, $c = 18$.

**Fig. 4.8** Recurrence plots (RP) and fuzzy recurrence plots (FRP) of state vectors of the phase-space reconstruction of the Rossler time series shown in Fig. 4.7, with embedding dimension = 3, and time delay = 1

#### 4.1.3.4 A Periodic Waveform

The sawtooth wave is a kind of non-sinusoidal waveform. The sawtooth function generates a sawtooth wave with a period of $2\pi$ for the elements of the time vector. Being like the sine of the elements of the time vector, it creates a sawtooth wave with peaks of +1 to −1 instead of a sine wave. Figure 4.9 shows a time series generated for 0.25 s of a 50 Hz sawtooth wave with a sample rate of 10 kHz.

For the construction of the RP, the threshold becomes negative if multiplying $\epsilon = 0.1$ or 0.2 with the mean of the signal. Therefore, $\epsilon$ is set as 0.1 or 0.2. RPs and FRPs of the sawtooth time series are shown in Fig. 4.10. Structural patterns of RPs and FRPs are quite similar in revealing the periodic recurrent behavior of the sawtooth signal by displaying diagonal lines with an equal distance. The uniformly gray small-scale texture of the FRPs of the sawtooth wave is quite visible in the form of lines parallel to the main diagonal of the FRPs. Such textural information can give more insight into the analysis of the underlying dynamics of the signal.

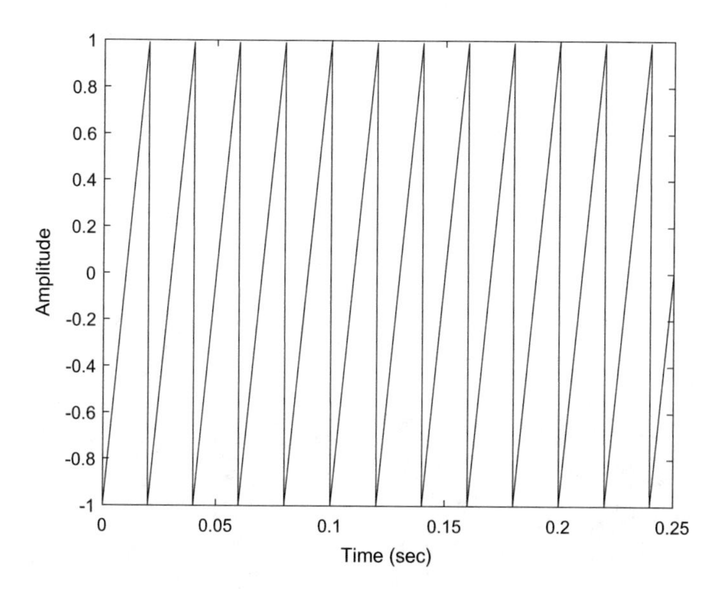


**Fig. 4.9** Time series of the sawtooth function

#### 4.1.3.5 An Aperiodic Waveform

The normalized cardinal sine function or normalized sinc function, denoted by sinc(x), is defined as

$$\text{sinc(x)} = \begin{cases} \frac{\sin(\pi x)}{\pi x} & : x \neq 0, \\ 1 & : x = 0. \end{cases} \tag{4.9}$$

The normalized sinc function is the Fourier transform of the rectangular function with no scaling. It is used for reconstructing a continuous bandlimited signal from uniformly spaced samples of that signal. Figure 4.11 shows a time series of 2000 points of the normalized sinc function.

Being like the previous illustrations, it can be appreciated that the FRPs provide much better visualization of the recurrence of the time series than the RPs by displaying good visual effects of checkerboard structures to reflect the aperiodic waveform (Fig. 4.12).

## 4.2 Fuzzy Cross Recurrence Plots

Let $\mathbf{X} = (\mathbf{x}_1, \ldots, \mathbf{x}_N)$ and $\mathbf{Y} = (\mathbf{y}_1, \ldots, \mathbf{y}_M)$ be the collections of the state vectors of the phase-space reconstructions of two time series, and $\mathbf{V}_X = \{\mathbf{v}_1(X), \ldots, \mathbf{v}_c(X)\}$ and $\mathbf{V}_\mathbf{Y} = \{\mathbf{v}_1(\mathbf{Y}), \ldots, \mathbf{v}_c(\mathbf{Y})\}$, where $c$ is the number of clusters, be the sets of dis-

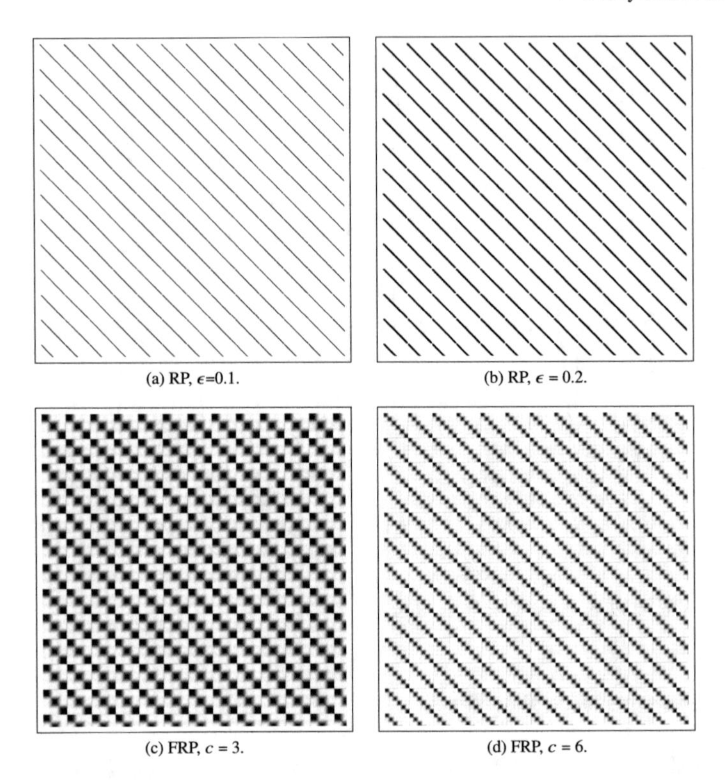


**Fig. 4.10** Recurrence plots (RP) and fuzzy recurrence plots (FRP) of state vectors of the phase-space reconstruction of the time series of the sawtooth function shown in Fig. 4.9, with embedding dimension = 3, and time delay = 1

tinct fuzzy cluster centers of **X** and **Y**, respectively. A fuzzy cross recurrence plot (FCRP) of **X** and **Y**, denoted as $C_{\mathbf{X}Y}$, can be expressed as the conjunction in fuzzy logic defined by the minimum T-norm (triangular norm) as [8]

$$\begin{aligned} C_{\mathbf{X}Y} &= \top_{\min}[\mathbf{U}(\mathbf{X}, \mathbf{Y}|\mathbf{V_X}), \mathbf{U}(\mathbf{X}, \mathbf{Y}|\mathbf{V_Y})] \\ &= \mu_{\mathbf{X}}(\mathbf{x}_i, \mathbf{y}_j) \wedge \mu_{\mathbf{Y}}(\mathbf{x}_i, \mathbf{y}_j), \\ &\quad i = 1, \ldots, N, j = 1, \ldots, M, \end{aligned} \tag{4.10}$$

where $\mathbf{U}(\mathbf{X}, \mathbf{Y}|\mathbf{V_X})$ and $\mathbf{U}(\mathbf{X}, \mathbf{Y}|\mathbf{V_Y})$ are fuzzy membership matrices of **X** and **Y** derived with reference to $\mathbf{V_X}$ and $\mathbf{V_Y}$, respectively, $\mu_{\mathbf{X}}(\mathbf{x}_i, \mathbf{y}_j)$ and $\mu_{\mathbf{Y}}(\mathbf{x}_i, \mathbf{y}_j)$ are fuzzy membership grades of similarity between $\mathbf{x}_i$ and $\mathbf{y}_j$ with reference to $\mathbf{V_X}$ and $\mathbf{V_Y}$, respectively, and $\wedge$ stands for the minimum operator.

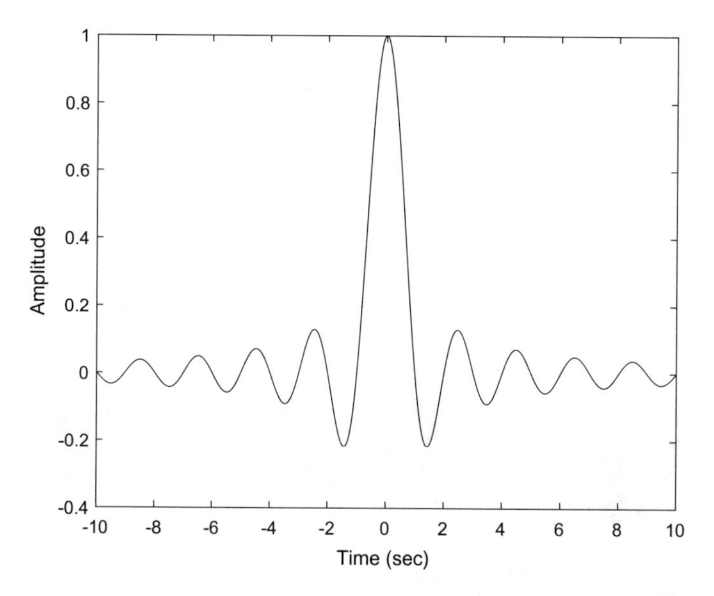


**Fig. 4.11** Time series of the normalized sinc function

The fuzzy membership function $\mu_{\mathbf{X}}(\mathbf{x}_i, \mathbf{y}_j) \in [0, 1]$, which expresses the relation or similarity between the two state vectors, can be inferred using two properties of the fuzzy relation [4] as follows:

1. Reflexivity:
   $\mu_{\mathbf{X}}(\mathbf{x}_i, \mathbf{y}_j) = 1$ if $\mathbf{x}_i = \mathbf{y}_j, i = 1, \ldots, N, j = 1, \ldots, M$.
2. Transitivity:
   $\mu_{\mathbf{X}}(\mathbf{x}_i, \mathbf{y}_j) = \max[\min\{\mu(\mathbf{x}_i, \mathbf{v}_k(\mathbf{X})), \mu(\mathbf{y}_j, \mathbf{v}_k(\mathbf{X}))\}], \quad k = 1, \ldots, c,$ which is called the max-min composition.

The fuzzy membership of $\mathbf{x}_i$ assigned to a cluster center $\mathbf{v}_k$ of $\mathbf{X}$, $\mu(\mathbf{x}_i, \mathbf{v}_k(\mathbf{X}))$ or denoted as $\mu_{ik}$, is computed using the FCM algorithm. The fuzzy membership function $\mu_{\mathbf{Y}}(\mathbf{x}_i, \mathbf{y}_j) \in [0, 1]$ can be obtained using the same fuzzy inference applied to $\mu_{\mathbf{X}}(\mathbf{x}_i, \mathbf{y}_j)$.

Finally, an FCRP, which is either a square or non-square plot, can be visualized as a grayscale image in the range [0, 1] by taking the complement of $C_{\mathbf{X}Y}$ that displays a black pixel if $\mathbf{x}_i = \mathbf{y}_j, i = 1, \ldots, N, j = 1, \ldots, M$, otherwise a pixel with a shade of gray.

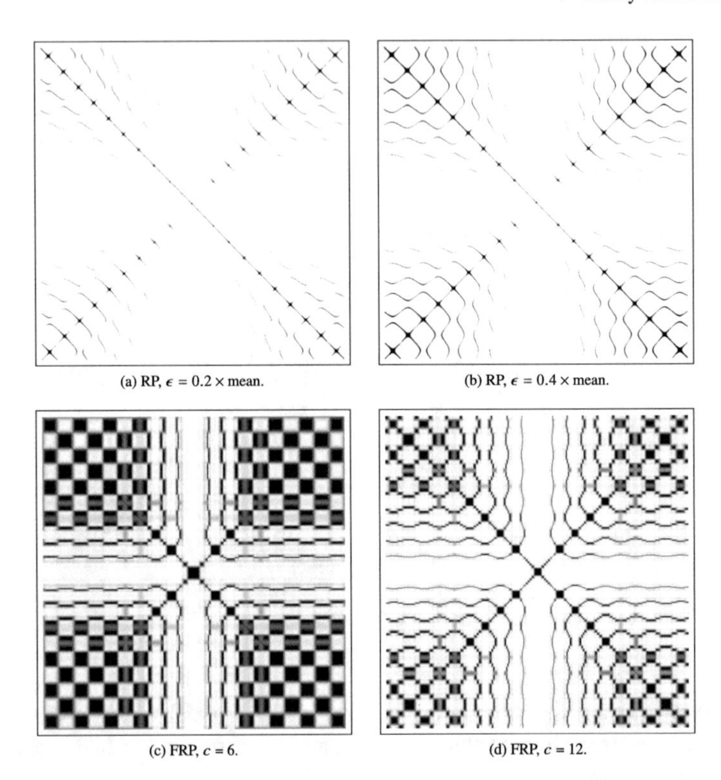
(a) RP, $\epsilon = 0.2 \times$ mean. (b) RP, $\epsilon = 0.4 \times$ mean.

(c) FRP, $c = 6$. (d) FRP, $c = 12$.

**Fig. 4.12** Recurrence plots (RP) and fuzzy recurrence plots (FRP) of state vectors of the phase-space reconstruction of the time series of the normalized sinc function shown in Fig. 4.11, with embedding dimension = 3, and time delay = 1

## 4.2.1 Computer Code

### Matlab Program for Constructing an FCRP

```
function B = fcrp(x1,x2,dim,tau,numclusters)
%-------------------------------------------------------------------------
% Fuzzy cross recurrence plot
% Ref: Tuan D. Pham
%-------------------------------------------------------------------------
% Input:
%       x1, x2: two time series
%       dim: embedding dimension (default value = 3)
%       tau: time delay (default value = 1)
%       numclusters: number of clusters (default value = 2)
% Output:
%       B: Fuzzy cross recurrence plot
```

```
% Other functions called:
%       pre_fcrp.m, getmu.m
%-----------------------------------------------------------------


% Fuzzy exponent for FCM
m=2;

% Compute phase-space vectors, find max centers, max fuzzy membership
[PS1,C1,U1] = pre_fcrp(x1,dim,tau,numclusters);
[PS2,C2,U2] = pre_fcrp(x2,dim,tau,numclusters);

% Compute fuzzy membership of PS1 wrt PS2 clusters for inference
for k=1:numclusters
    for j=1:length(PS1)
        U12(k,j) = getmu(PS1(j,:),C2(k,:),C2,m);
    end
end

% Compute fuzzy membership of PS2 wrt PS1 clusters for inference
for k=1:numclusters
    for j=1:length(PS2)
        U21(k,j) = getmu(PS2(j,:),C1(k,:),C1,m);
    end
end

% Use min-operator for P1
P1=zeros(length(PS1),length(PS2),numclusters);
for k=1:numclusters
    for i=1:length(PS1)
        for j=1:length(PS2)
            if norm(PS1(i,:) - PS2(j,:)) == 0
                P1(i,j,k)= 1;
            elseif U1(k,i)>= U21(k,j)
                P1(i,j,k) = U21(k,j);
            else
                P1(i,j,k) = U1(k,i);
            end
        end
    end
end

% Use max-operator for A1
A1=zeros(length(PS1),length(PS2));
for k=1:numclusters
    for i=1:length(PS1)
        for j=1:length(PS2)
            if k==1
                A1(i,j)=P1(i,j,k);
            else
                if P1(i,j,k)>=P1(i,j,k-1)
                    A1(i,j)=P1(i,j,k);
                end
            end
        end
    end
```

```
end


% Use min-operator for P2
P2=zeros(length(PS1),length(PS2),numclusters);
for k=1:numclusters
    for j=1:length(PS2)
        for i=1:length(PS1)
            if norm(PS2(j,:) - PS1(i,:)) == 0
                P2(i,j,k)= 1;
            elseif U2(k,j)>= U12(k,i)
                P2(i,j,k) = U12(k,i);
            else
                P2(i,j,k) = U2(k,j);
            end
        end
    end
end

% Use max-operator for A2
A2=zeros(length(PS1),length(PS2));
for k=1:numclusters
    for j=1:length(PS2)
        for i=1:length(PS1)
            if k==1
                A2(i,j)=P2(i,j,k);
            else
                if P2(i,j,k)>=P2(i,j,k-1)
                    A2(i,j)=P2(i,j,k);
                end
            end
        end
    end
end

% Combine A1 AND A2 with fuzzy operator conjunction "AND"
B = min(A1,A2);

% Making dark pixels for recurrences
B=imcomplement(B);
figure
imshow(B)

%-----------------------------------------------------
function [PS,Center,U] = pre_fcrp(x,dim,tau,cluster)
%-----------------------------------------------------
% Preprocessing of fuzzy cross recurrence plot
% Reference: Tuan D. Pham
%----------------------------------------------------
% Input:
%       x: time series
%       dim: embedding dimension (default value = 3)
%       tau: time delay (default value = 1)
%       cluster: number of clusters (default value = 2)
% Output:
%       PS = state vectors
```

```
%       C = clusters of max membership
%       U = max memberships
%-------------------------------------------------------


[nRow,~] = size(x);

if (nRow==1)
    x = x';
    [nRow,~] = size(x);
end

M=nRow-(dim-1)*tau;
PS=zeros(M,dim);

% Construct state vectors
for i=1:M
    for j=1:dim
        PS(i,j)=x(i+(j-1)*tau);
    end
end

% Compute fuzzy c-means using Matlab toolbox
[Center, U, ~] = fcm(PS,cluster);

%-----------------------------------
function mu = getmu(x,v,C,m)
%-----------------------------------
% Compute fuzzy membership from FCM
% Ref: Tuan D. Pham
%-----------------------------------

dxv = norm(x-v);
for i = 1:length(C)
    d(i)=norm(x-C(i,:));
end

B = dxv/sum(d);
mu = 1/(B^(2/(m-1)));
```

## Matlab Program for Demonstration of Computing FCRPs

```
% Demonstration of computing fuzzy cross RPs

% Input parameters
dim=3;
tau=1;
numclusters = 6;
```

```
rng('default');

% Lorenz components
L = textread('Lorenz.txt');
Lx = L(1:400,2);
Ly = L(1:400,3);
Lz = L(1:400,4);
%Lx = normalize(Lx,'range'); % Scale in interval [0,1]
%Ly = normalize(Ly,'range');
%Lz = normalize(Lz,'range');

% Fuzzy Cross RPs
Cxy = fcrp(Lx,Ly,dim,tau,numclusters);
imwrite(Cxy,'Cxy-frp.jpg');

Cxz = fcrp(Lx,Lz,dim,tau,numclusters);
imwrite(Cxz,'Cxz-frp.jpg');

Cyz = fcrp(Ly,Lz,dim,tau,numclusters);
imwrite(Cyz,'Cyz-frp.jpg');
```

---

### 4.2.2 Illustrations

Figure 4.13 shows the CRPs and FCRPs of the time series of 4000 data points of the three components of the Lorenz system, using embedding dimension = 3, and time delay = 1 for the phase-space reconstructions. The similarity threshold used for constructing the CRPs are set to be 0.2. The number of clusters for computing the FCRPs are 6.

The recurrence the underlying dynamics of a system is characterized by typical patterns: (1) a homogeneous distribution of points indicates that the system is associated with stationary stochastic processes, for example, Gaussian white noise; (2) long diagonal lines have an implication of periodic behaviors; and (3) white areas or bands mean a nonstationary process and abrupt changes in the system dynamics, such as chaos.

The CRPs of the pairs of the Lorenz components show white areas in circles, while the FCRPs show white rectangular blocks of texture. Both CRPs and FCRPs indicate the chaotic patterns of the Lorenz system. Regarding the analysis of the recurrence quantification, such blocks of texture provided by the FCRP method would be more useful for the purpose of developing quantitative measures of the recurrence dynamics.

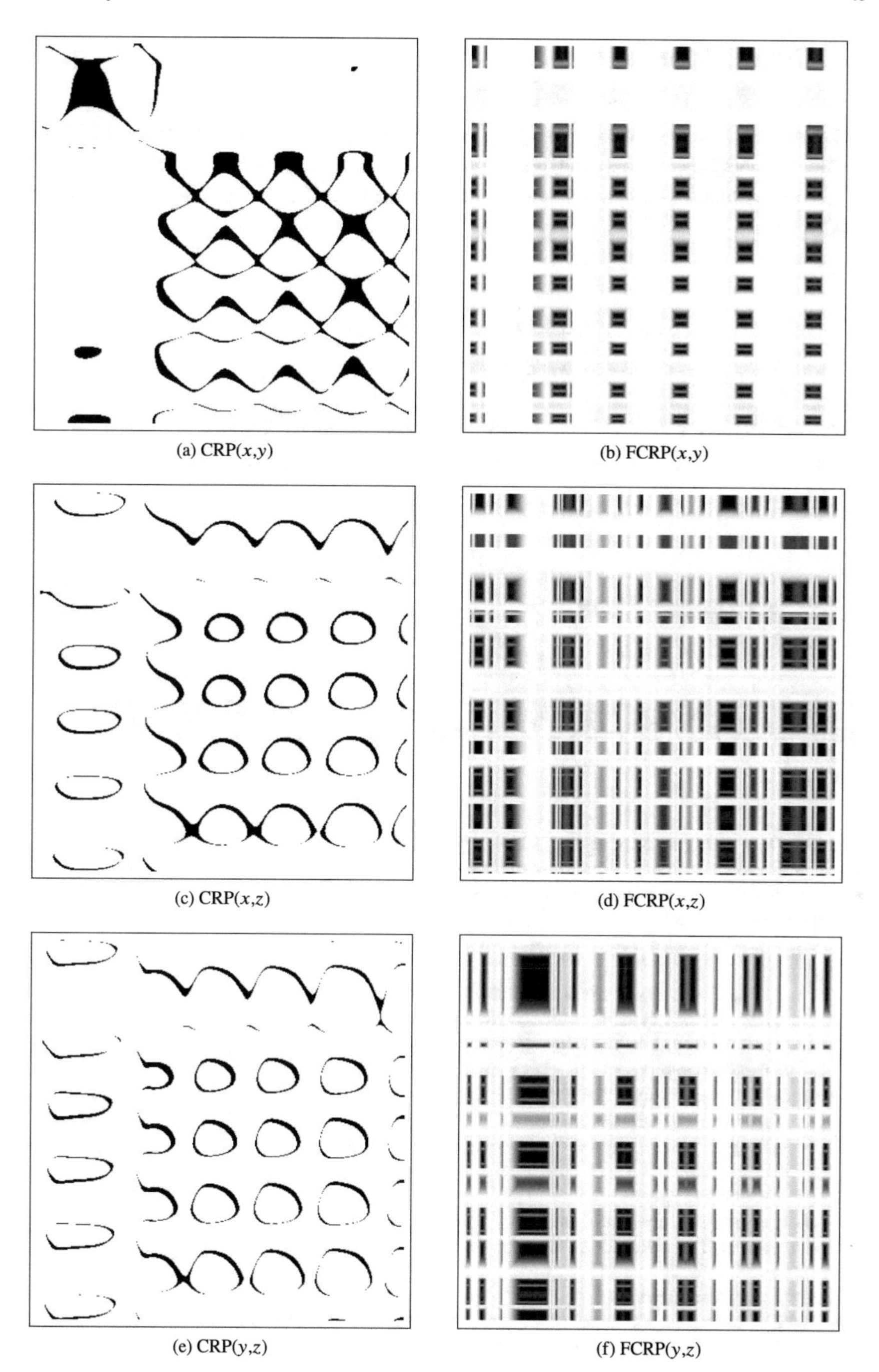


**Fig. 4.13** Cross recurrences of state vectors of the phase-space reconstructions of time series of three Lorenz components $x$, $y$, and $z$, with embedding dimension = 3, and time delay = 1: (a), (c), and (e) are cross recurrence plots with similarity threshold = 0.2; and (b), (d), and (f) are fuzzy cross recurrence plots with the number of clusters = 5

## 4.3 Fuzzy Joint Recurrence Plots

Let $\mathbf{X} = (\mathbf{x}_1, \ldots, \mathbf{x}_N)$ and $\mathbf{Y} = (\mathbf{y}_1, \ldots, \mathbf{y}_N)$ be the collections of the state vectors of the phase-space reconstructions of two time series. A fuzzy joint recurrence plot (FJRP) of $\mathbf{X}$ and $\mathbf{Y}$, denoted as $J_{\mathbf{X}Y}$, can be defined as the product T-norm of the two FRPs [8]:

$$\begin{aligned} J_{\mathbf{X}Y} &= \top_{\text{prod}}[\text{FRP}(\mathbf{X}), \text{FRP}(\mathbf{Y})] \\ &= \tilde{\mathbf{R}}(\mathbf{x}_i, \mathbf{x}_j) \cdot \tilde{\mathbf{R}}(\mathbf{y}_i, \mathbf{y}_j), i, j = 1, \ldots, N, \end{aligned} \tag{4.11}$$

where FRP($\mathbf{X}$) and FRP($\mathbf{Y}$) are fuzzy recurrence plots of $\mathbf{X}$ and $\mathbf{Y}$, respectively, and the notation $\cdot$ stands for the ordinary product of real numbers. As a result, $J_{\mathbf{X}Y}$ is a square plot.

### *4.3.1 Computer Code*

**Matlab Program for Constructing an FJRP**

```
function P = fjrp(x1,x2,dim,tau,numclusters)
%------------------------------------------------------------------------
% Fuzzy joint recurrence plot of two time series
% Ref: Tuan D. Pham
%------------------------------------------------------------------------
% Input:
%       x1, x2: two time series of same length
%       dim: embedding dimension (default value = 3)
%       tau: time delay (default value = 1)
%       cluster: number of clusters (default value = 2)
% Output: P (fuzzy joint recurrence plot)
% Other function called: frp_mu.m
%-----------------------------------------------------------------

% Fuzzy recurrence plots
P1 = frp_mu(x1,dim,tau,numclusters);
P2 = frp_mu(x2,dim,tau,numclusters);

% Fuzzy joint recurrence plot
P = P1.*P2; % product t-norm (ordinary product of real numbers)
P=imcomplement(P); % making dark pixels for recurrences

figure
imshow(P)
```

```
%------------------------------------------------------------
function FRP = frp_mu(x,dim,tau,cluster)
%------------------------------------------------------------
% Input:
%       x: time series
%       dim: embedding dimension  (default value = 3)
%       tau: time delay (default value = 1)
%       cluster: number of clusters (default value = 2)
% Output:
%       FRP: Fuzzy recurrence plot
%------------------------------------------------------------
rng('default') % set random number genenator to default values

switch nargin
    case 1
        dim=3;
        tau=1;
        cluster=2;
        T=NaN;
    case 2
        tau=1;
        cluster=2;
        T=NaN;
    case 3
        cluster=2;
        T=NaN;
    case 4
        T=NaN;
end

[nRow,~] = size(x);

if (nRow==1)
    x = x';
    [nRow,~] = size(x);
end

M=nRow-(dim-1)*tau;
PS=zeros(M,dim);

for i=1:M
    for j=1:dim
        PS(i,j)=x(i+(j-1)*tau);
    end
end



[~, FR, ~] = fcm(PS,cluster); % use fuzzy c-means from Matlab toolbox

cRP=zeros(length(FR),length(FR),cluster);

for c=1:cluster
    for i=1:length(FR)
        for j=1:length(FR)
```

```
            if norm(PS(i,:) - PS(j,:)) == 0
                cRP(i,j,c)= 1; % reflexive
            elseif FR(c,i)>=FR(c,j)
                cRP(i,j,c)=FR(c,j);
            else
                cRP(i,j,c)=FR(c,i);
            end
        end
    end
end

FRP=zeros(length(cRP),length(cRP));
for c=1:cluster
    for i=1:length(cRP)
        for j=1:length(cRP)
            if c==1
                FRP(i,j)=cRP(i,j,c);
            else
                if cRP(i,j,c)>=FRP(i,j)
                    FRP(i,j)=cRP(i,j,c);
                end
            end
        end
    end
end

end
```

## Matlab Program for Demonstration of Computing FJRPs

```
% Demonstration of computing fuzzy joint RPs

% Input parameters
dim=3;
tau=1;
numclusters = 6;
rng('default');

% Lorenz components
L=textread('Lorenz.txt');
Lx = L(1:400,2);
Ly = L(1:400,3);
Lz = L(1:400,4);
%Lx = normalize(Lx,'range'); % Scale in interval [0,1]
%Ly = normalize(Ly,'range');
%Lz = normalize(Lz,'range');

% Fuzzy Joint RPs
Jxy = fjrp(Lx,Ly,dim,tau,numclusters);
```

```
imwrite(Jxy,'Jxy-frp.jpg');

Jxz = fjrp(Lx,Lz,dim,tau,numclusters);
imwrite(Jxz,'Jxz-frp.jpg');

Jyz = fjrp(Ly,Lz,dim,tau,numclusters);
imwrite(Jyz,'Jyz-frp.jpg');
```

### 4.3.2 *Illustrations*

Figure 4.14 shows the JRPs and FJRPs of the time series of 4000 data points of the three components of the Lorenz system, using embedding dimension = 3, and time delay = 1 for the phase-space reconstructions. The similarity threshold used for constructing the JRPs are set to be 0.2. The number of clusters for computing the FJRPs is 6.

Both JRPs and FJRPs show similar patterns of the joint recurrences of the pair-wise components of the Lorenz system, including large white areas and diagonal lines. However, the FJRPs display more visual information and texture of the recurrences. The JRPs and FJRPs show the time points at which a recurrence in one component of the dynamical system occurs simultaneously with a recurrence of another component of the same system. These plots express both chaotic and periodical behaviors of the joint recurrences.

## 4.4 Texture Analysis of Fuzzy Recurrence Plots

Because FRPs display the recurrences of nonlinear system dynamics with textural information, many texture analysis methods developed in the literature of image processing can be applied to quantifying their patterns in order to gain more insights into the underlying dynamics and changing behaviors of such complex systems. Some methods for texture analysis for potential applications to characterizing different complex visual displays of FRPs are as follows:

1. The gray-level co-occurrence matrix (GLCM) [9].
2. The semivariogram [10].
3. Local binary patterns (LBP) [11].
4. Texture spectrum [12].
5. Wavelets [13].
6. Fuzzy Kolmogorov–Sinai entropy [14].
7. Fractal analysis [15].

Some reviews and literature on texture analysis methods can be found in [16–18].

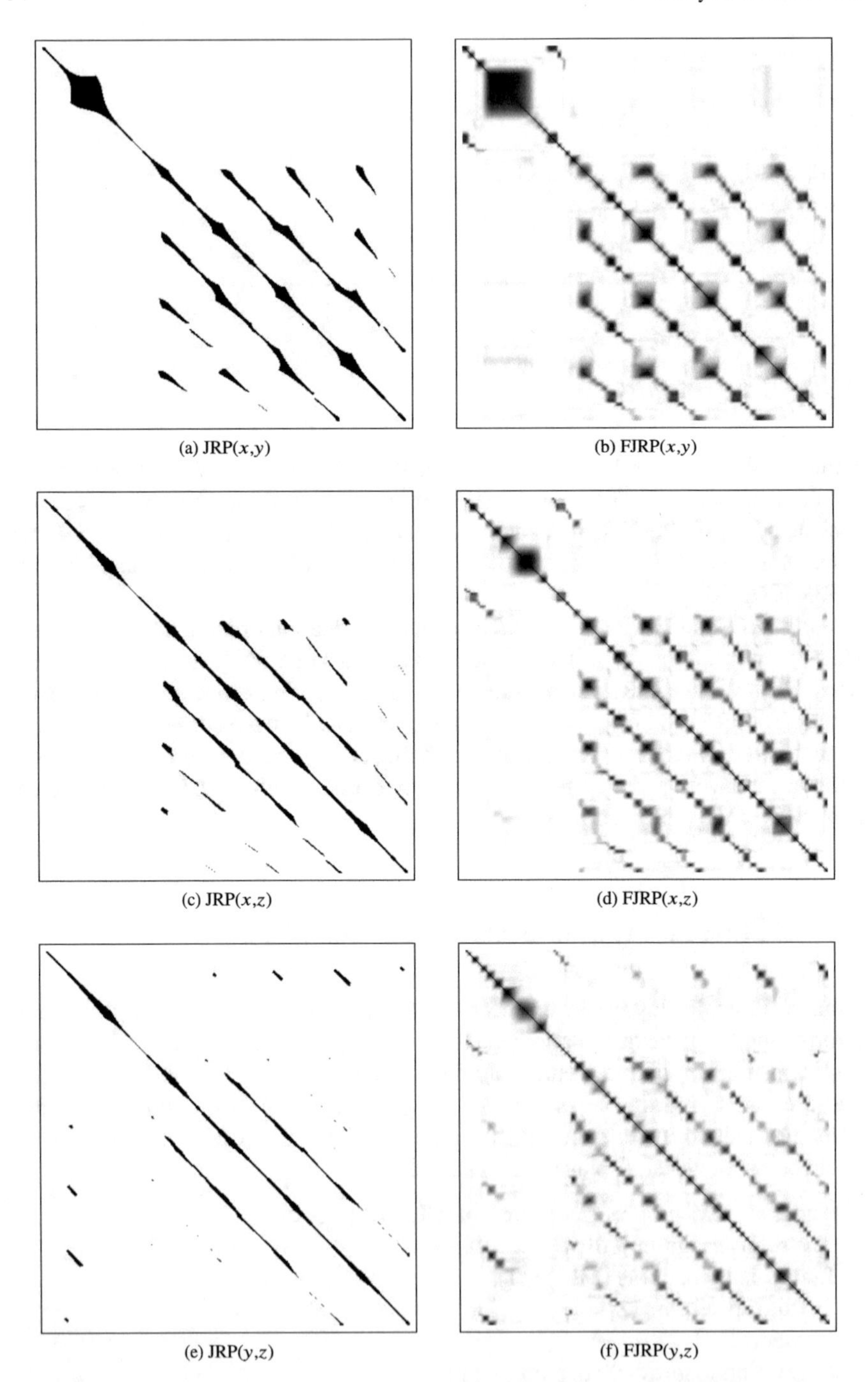


**Fig. 4.14** Joint recurrences of state vectors of the phase-space reconstructions of time series of three Lorenz components $x$, $y$, and $z$, with embedding dimension = 3, and time delay = 1: **a**, **c**, and **e** are joint recurrence plots with similarity threshold = 0.2; and **b**, **d**, and **f** are fuzzy joint recurrence plots with the number of clusters = 6.

## References

1. Bezdek JC (1981) Pattern recognition with fuzzy objective function algorithms. Plenum Press, New York
2. Bezdek JC, Ehrlich R, Full W (1984) FCM: the fuzzy c-means clustering algorithm. Comput Geosci 10:191–203
3. Pham TD (2016) Fuzzy recurrence plots. EPL (Eur Lett) 116:50008
4. Zadeh LA (1971) Similarity relations and fuzzy orderings. Inf Sci 3:177–200
5. Lorenz EN (1963) Deterministic nonperiodic flow. J Atmos Sci 20:130–141
6. Glass L, Mackey M (2010) Mackey-Glass equation. Scholarpedia 5:6908. https://doi.org/10.4249/scholarpedia.6908
7. Rossler OE (1976) An equation for continuous chaos. Phys Lett A 57:397–398
8. Pham TD (2020) Fuzzy cross and fuzzy joint recurrence plots. Physica A 540:123026
9. Haralick RM, Shanmugam K, Dinstein I (1973) Textural features for image classification. IEEE Trans Syst, Man Cybern 3:610–621
10. Pham TD (2016) The semi-variogram and spectral distortion measures for image texture retrieval. IEEE Trans Image Process 25:1556–1565
11. Ojala T, Pietikainen M, Harwood D (1996) A comparative study of texture measures with classification based on feature distributions. Pattern Recognit 29:51–59
12. He DC, Wang L (1990) Texture unit, texture spectrum, and texture analysis. IEEE Trans Geosci Remote Sens 28:509–512
13. Wouwer GVD, Scheunders F, Dyck DV (1999) Statistical texture characterization from discrete wavelet representation. IEEE Trans Image Process 8:592–598
14. Pham TD (2016) The Kolmogorov-Sinai entropy in the setting of fuzzy sets for image texture analysis and classification. Pattern Recognit 53:229–237
15. Kaplan LM (1999) Extended fractal analysis for texture classification and segmentation. IEEE Trans Image Process 8:1572–1585
16. Zhang J, Tan T (2002) Brief review of invariant texture analysis methods. Pattern Recognit 35:735–747
17. Petrou MMP, Sevilla PG (2006) Image processing: dealing with texture. Wiley, West Sussex
18. Hung CC, Song E, Lan Y (2019) Image texture analysis. Springer Nature, Switzerland

# Chapter 5
# Fuzzy Recurrence Networks

## 5.1 Unweighted Fuzzy Recurrence Networks

Let $\alpha \in [0, 1]$ be a constant fuzzy membership grade. Based on the definition of the adjacency matrix of an undirected and unweighted recurrence network [1], the adjacency matrix of an undirected and unweighted fuzzy recurrence network at an $\alpha$ cut, which is a defuzzified (binary) adjacency matrix and denoted as $\mathbf{G}_\alpha$, can be defined in terms of an FRP at an $\alpha$ cut, $\tilde{\mathbf{R}}_\alpha$, as [2]

$$\mathbf{G}_\alpha = \tilde{\mathbf{R}}_\alpha - \mathbf{I}, \tag{5.1}$$

where $\mathbf{I}$ is the identity matrix, and

$$\tilde{\mathbf{R}}_\alpha = \begin{cases} 1 : \tilde{\mathbf{R}}(i, j) \geq \alpha, \\ 0 : \text{otherwise}, \end{cases} \tag{5.2}$$

where $\tilde{\mathbf{R}}(i, j), i, j = 1, \ldots, N$, is an element of FRP $\tilde{\mathbf{R}}$.

As a result, a defuzzified recurrence network is a binary recurrence network that is constructed from the $\alpha$-cut FRP, which no longer holds the fuzzy membership values between 0 and 1.

The need for scalability of a fuzzy recurrence network (FRN) is addressed as follows. Let $\tau$ and $m$ be the time delay and embedding dimension, respectively. The number of the state vectors of a dynamical system of $N$-point time series is $G = N - (m - 1)\tau$. It is obvious that, for a very large value of $N$, a recurrence network will have a corresponding very large number of vertices that is proportional to $G$ if the adjacency matrix is not sparse. Such a situation makes it computationally expensive or impractical for computing the recurrence network properties with respect to the computational complexity.



T. D. Pham, *Fuzzy Recurrence Plots and Networks with Applications in Biomedicine*, https://doi.org/10.1007/978-3-030-37530-0_5

The scalability of a defuzzified adjacency matrix of an undirected $\alpha$-cut recurrence network can be obtained by replacing the state vectors with a number of prototypes of the state vectors, denoted as $c$, that are the number of fuzzy cluster centers and much smaller than the number of the state vectors used for constructing $G$.

Let $\mathbf{X} = (\mathbf{x}_1, \ldots, \mathbf{x}_N)$ be a collection of the state vectors of the reconstructed phase space of a time series, $\mathbf{V} = \{\mathbf{v}_1, \ldots, \mathbf{v}_c\}$ be a set of clusters, and the fuzzy membership grades expressing the degrees of the state vectors $\mathbf{x}_i \in \mathbf{X}$ belonging to cluster centers $\mathbf{v}_k \in \mathbf{V}$, denoted as $\mu(\mathbf{x}_i, \mathbf{v}_k)$. Using the fuzzy relations, we can infer the fuzzy membership grades of similarity between cluster pairs as

- $\mu(\mathbf{v}_i, \mathbf{v}_i,) = 1, \ \forall \mathbf{v}_i \in \mathbf{V}$.
- $\mu(\mathbf{x}_i, \mathbf{v}_k) = \mu(\mathbf{v}_k, \mathbf{x}_i), \ \forall \mathbf{x}_i \in \mathbf{X}, \forall \mathbf{v}_k \in \mathbf{V}$.
- $\mu(\mathbf{v}_j, \mathbf{v}_k)) = \max[\min\{\mu(\mathbf{v}_j, \mathbf{x}_i), \mu(\mathbf{x}_i, \mathbf{v}_k)\}], \ i = 1, \ldots, N$.

A defuzzified $\beta$-cut prototype recurrence matrix of size $c \times c$ $(1 < c < N)$, denoted as $\mathbf{M}_\beta$, can be obtained by

$$\mathbf{M}_\beta = \begin{cases} 1 : \mu(\mathbf{v}_j, \mathbf{v}_k) \geq \beta \\ 0 : \text{otherwise,} \end{cases} \tag{5.3}$$

where $\beta \in [0, 1]$.

Finally, the adjacency matrix of an undirected and unweighted $\beta$-cut prototype recurrence network, denoted as $B_\beta$, is defined as follows:

$$\mathbf{B}_\beta = \mathbf{M}_\beta - \mathbf{I}. \tag{5.4}$$

### 5.1.1 Computer Code

**Matlab Program for Constructing an FRN**

```
function [FRP,A,B1,B2] = frNet(x,dim,tau,ncluster,alpha,beta)
%---------------------------------------------------------------------------
% Reference: Pham TD (2017)
% From fuzzy recurrence plots to scalable recurrence networks of time series,
% EPL 118: 20003.
%---------------------------------------------------------------------------
% Input:
%        x: time series
%        dim: embedding dimension
%        tau: time delay
%        ncluster: number of clusters
%        alpha: cuttoff fuzzy membership threshold.
%
% Output:
%        FRP: Defuzzified recurrence plot
%        A: Defuzzified adjacency matrix using alpha-cut
%        B1: fuzzy recurrence prototype matrix/plot
```

```
%       B2: Defuzzified prototype adjacency matrix using beta-cut
% Test:
%       x = randi([0 5],1,500); dim=3; tau=1; ncluster=3;
%       alpha=0.5; beta=0.02;
%       [FRP,A,B1,B2]=frNet(x,dim,tau,ncluster,alpha,beta);
%-------------------------------------------------------------


[nRow,~] = size(x);

if (nRow==1)
    x = x';
    [nRow,~] = size(x);
end

M=nRow-(dim-1)*tau;
PS=zeros(M,dim);

for i=1:M
    for j=1:dim
        PS(i,j)=x(i+(j-1)*tau);
    end
end

[center,FR,~] = fcm(PS, ncluster); % use FCM from Matlab Toolbox

cRP=zeros(length(FR),length(FR),ncluster);

for c=1:ncluster
    for i=1:length(FR)
        for j=1:length(FR)
            if norm(PS(i,:) - PS(j,:)) == 0
                cRP(i,j,c)= 1; % reflexive
            elseif FR(c,i)>=FR(c,j)
                cRP(i,j,c)=FR(c,j);
            else
                cRP(i,j,c)=FR(c,i);
            end
        end
    end
end

FRP=zeros(length(cRP),length(cRP));
for c=1:ncluster
    for i=1:length(cRP)
        for j=1:length(cRP)
            if c==1
                FRP(i,j)=cRP(i,j,c);
            else
                if cRP(i,j,c)>=FRP(i,j)
                    FRP(i,j)=cRP(i,j,c);
                end
            end
        end
    end
end

% fuzzy recurrence plots (frp)
frp = imcomplement(FRP);
imshow(frp)
%print('frp', '-deps', '-r300');
figure
```

```
% defuzzified recurrence plot (FRP)
FRP(FRP>=alpha)=1;
FRP(FRP<alpha)=0;

FRP=imcomplement(FRP); % making dark pixels for recurrences
imshow(FRP)

% Defuzzified adjacency matrix of state-space phases
I = eye(length(FRP)); % identity matrix
A = imcomplement(FRP)-I; % recurrence adjacency matrix

% Defuzzified adjacency matrix of prototypes
Y=zeros(ncluster,ncluster,length(FR));

for c=1:length(FR)
    for i=1:ncluster
        for j=1:ncluster
            if norm(center(i,:) - center(j,:)) == 0
                Y(i,j,c)= 1; % reflexive
            elseif FR(i,c)>=FR(j,c)
                Y(i,j,c)=FR(j,c);
            else
                Y(i,j,c)=FR(i,c);
            end
        end
    end
end

B1 = zeros(ncluster,ncluster);

for c=1:length(FR)
    for i=1:ncluster
        for j=1:ncluster
            if c==1
                B1(i,j)=Y(i,j,c);
            else
                if Y(i,j,c)>=B1(i,j)
                    B1(i,j)=Y(i,j,c);
                end
            end
        end
    end
end

B2=B1;
B2(B2>=beta)=1;
B2(B2<beta)=0;

% Defuzzified adjacency matrix of prototypes of state vectors
I = eye(length(B2)); % identity matrix
B2 = B2-I; % prototype recurrence adjacency matrix

% plot graph
G=graph(B2);
figure
plot(G)
```

### 5.1.2 Illustrations

Figure 5.1 shows the unweighted fuzzy recurrence networks of $x$-, $y$-, and $z$-components of the Lorenz system. The length of the time series of each component is 4000 time points. The phase space is reconstructed using embedding dimension = 3 and time delay = 1. Two networks for each Lorenz-system component were obtained using $\alpha = 0.5$ with two different values for $\beta = 0.1$ and 0.3.

As being expected, the network topology of the three components are denser using a smaller cutoff value of $\beta$ (0.1), allowing more nodes of the networks to relate to each other. Constructing the networks using various cutoff membership grades and determination of their corresponding graph properties can offer as a useful tool for understanding complex patterns of the dynamical system.

As another example, FWRNs were constructed by using a publicly available PhysioBank dataset that includes electromyograms (EMG) recordings from three subjects: one is without neuromuscular disease (healthy), one with myopathy, and one with neuropathy [3]. The length of the three EMG time series is 5000 data points, which are shown in Fig. 5.2. The embedding dimension and time delay chosen for the FWRN were 3 and 1, respectively.

Being similar to the case of the Lorenz system, all networks of the healthy, myopathy, and neuropathy subjects have larger degrees of nodes (the number of edges connected to each node) with the smaller value of $\beta$ (0.2), while the degrees of nodes of the network using $\beta = 0.5$ are zero. It can be observed that, for $\beta = 0.2$, the network topology of the healthy is different from those of the myopathy and neuropathy. The comparison of network topologies can provide valuable information for differentiating healthy control from disease and stages of the disorder (Fig. 5.3).

## 5.2 Fuzzy Weighted Recurrence Networks

Let $\mathbf{X} = \{\mathbf{x}\}$ be the set of phase-space states, $N$ a given number of clusters of the states, and a set of $N$ fuzzy clusters, $\mathbf{V} = \{\mathbf{v}_i : i = 1, \ldots, N\}$. Fuzzy clusters can be defined as groups that contain data points, where each data point has a degree of fuzzy membership of belonging to each group. Explanation about the concept and technical formulation of fuzzy clustering in terms of the fuzzy $c$-means algorithm has been described in Chap. 3. By analogy with the inference for constructing a fuzzy recurrence plot and scalable network [2], a fuzzy relation $\tilde{\mathbf{R}}$ between $\mathbf{v}_i$ and $\mathbf{v}_j$, $i, j = 1, \ldots, N$, is characterized by a fuzzy membership function $\mu \in [0, 1]$, which expresses the degree of similarity of each pair $(\mathbf{v}_i, \mathbf{v}_j)$ in $\tilde{\mathbf{R}}$, and has the following three properties [4]:

1. Reflexivity: $\mu(\mathbf{v}_i, \mathbf{v}_i) = 1, \ \forall \mathbf{v}_i \in \mathbf{V}$.
2. Symmetry: $\mu(\mathbf{v}_i, \mathbf{x}) = \mu(\mathbf{x}, \mathbf{v}_i), \ \forall \mathbf{x} \in \mathbf{X}, \forall \mathbf{v}_i \in \mathbf{V}$.
3. Transitivity: $\mu(\mathbf{v}_i, \mathbf{v}_j) = \max[\min\{\mu(\mathbf{v}_i, \mathbf{x}), \mu(\mathbf{v}_j, \mathbf{x})]\}, \ \forall \mathbf{x} \in \mathbf{X}, \forall \mathbf{v}_i, \mathbf{v}_j \in \mathbf{V}$.

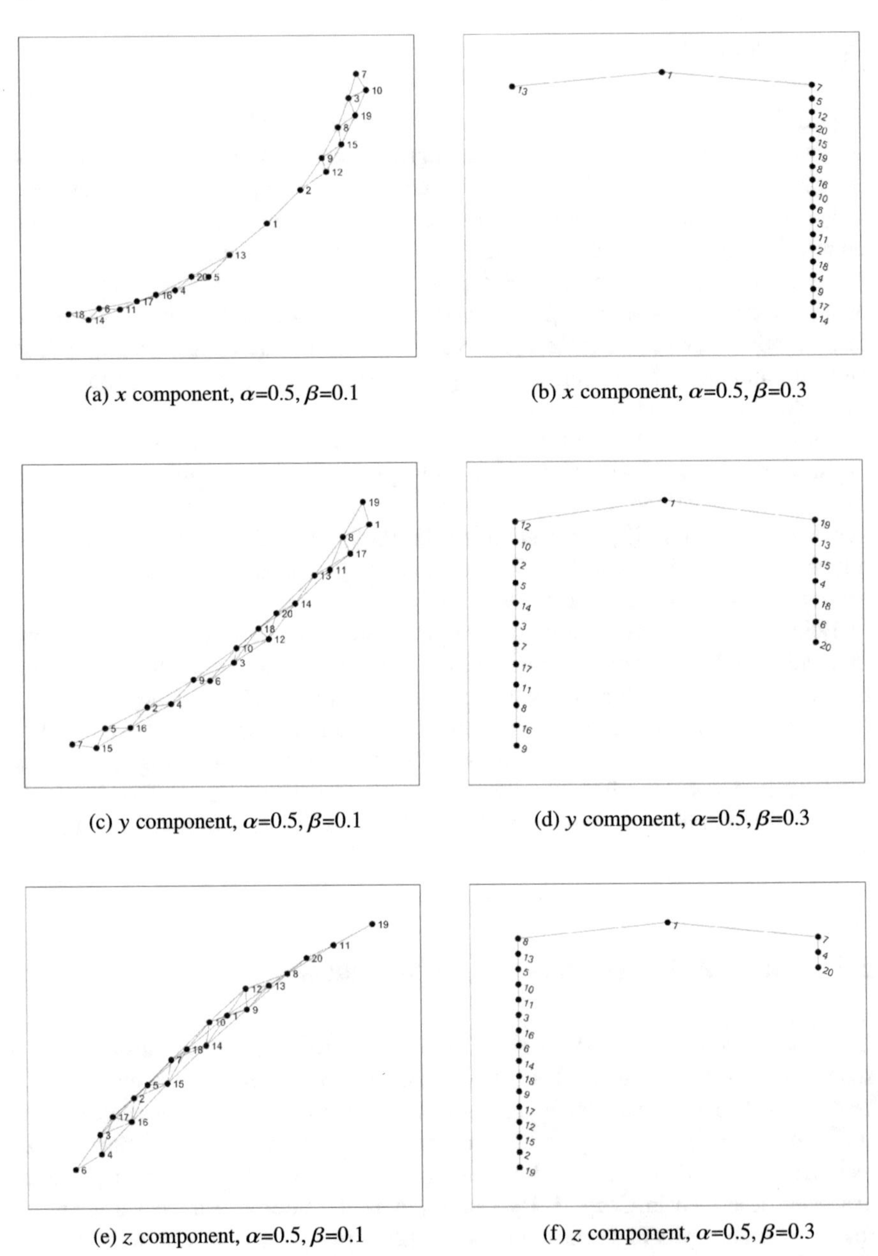

(a) $x$ component, $\alpha$=0.5, $\beta$=0.1

(b) $x$ component, $\alpha$=0.5, $\beta$=0.3

(c) $y$ component, $\alpha$=0.5, $\beta$=0.1

(d) $y$ component, $\alpha$=0.5, $\beta$=0.3

(e) $z$ component, $\alpha$=0.5, $\beta$=0.1

(f) $z$ component, $\alpha$=0.5, $\beta$=0.3

**Fig. 5.1** Scalable fuzzy recurrence networks of three Lorenz-system components, using embedding dimension = 3, time delay = 1, and number of clusters = 20 for the phase-space reconstruction of each time series of 4000 data points

**Fig. 5.2** Time series of 5000 data points of EMG signals

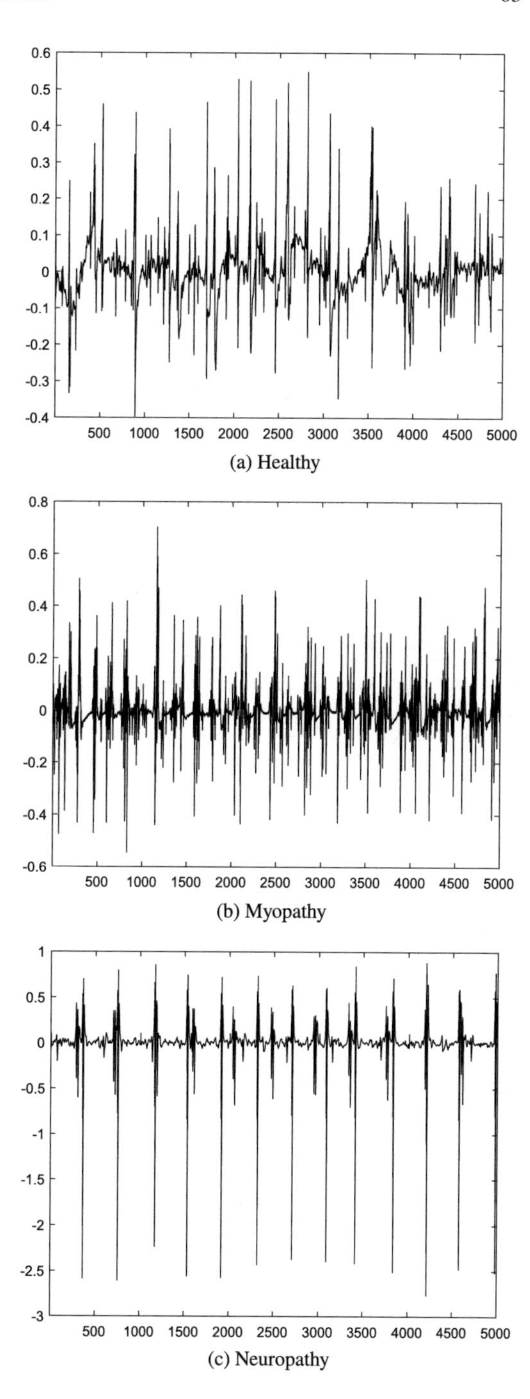


(a) Healthy

(b) Myopathy

(c) Neuropathy

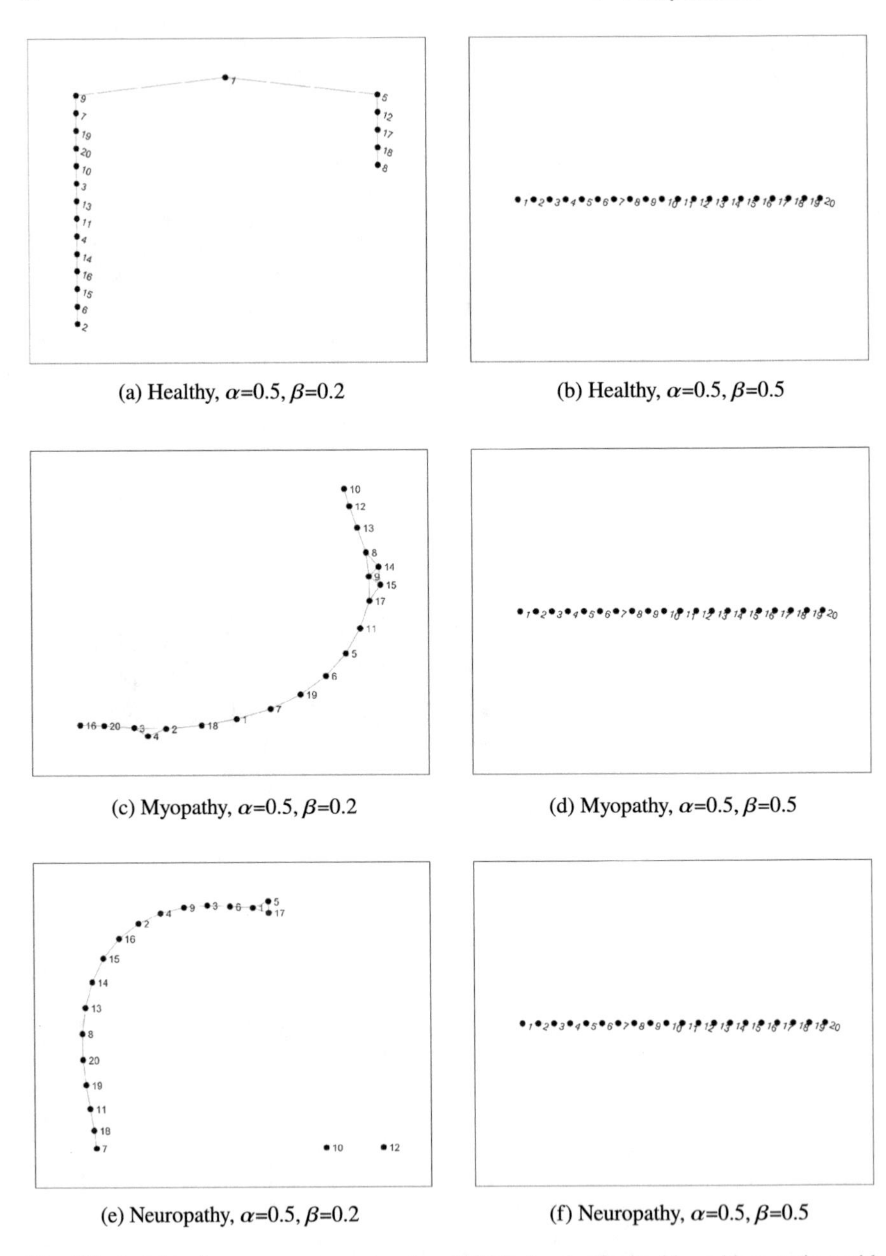


(a) Healthy, $\alpha$=0.5, $\beta$=0.2

(b) Healthy, $\alpha$=0.5, $\beta$=0.5

(c) Myopathy, $\alpha$=0.5, $\beta$=0.2

(d) Myopathy, $\alpha$=0.5, $\beta$=0.5

(e) Neuropathy, $\alpha$=0.5, $\beta$=0.2

(f) Neuropathy, $\alpha$=0.5, $\beta$=0.5

**Fig. 5.3** Scalable fuzzy recurrence networks of EMG signals of a healthy subject, patient with myopathy, and one with neuropathy, using embedding dimension = 3, time delay = 1, and number of clusters = 20 for the phase-space reconstruction of each time series of 5000 data points

An $N \times N$ fuzzy weighted recurrence network (FWRN) can be constructed with an associated fuzzy weighted adjacency matrix as

$$\mathbf{W} = \tilde{\mathbf{R}} - \mathbf{I}, \quad (5.5)$$

where $\mathbf{W}$ is an $N \times N$ adjacency matrix of edge weights, and $\mathbf{I}$ is the $N \times N$ identity matrix.

### 5.2.1 Computer Code

**Matlab Program for Constructing an FWRN**

```
function [wFRN,wPFRN] = fWrNet(x,dim,tau,ncluster)
%------------------------------------------------------------------------
% Fuzzzy weighted recurrence networks
% Reference: Pham TD (2019)
% Fuzzy weighted recurrence networks of time series,
% Physica A 513: 409-417.
%------------------------------------------------------------------------
% Input:
%       x: time series
%       dim: embedding dimension of a reconstructed phase space
%       tau: time delay to reconstruct phase space
%       ncluster: number of clusters for fuzzy c-means clustering
%
% Output:
%       wFRN: fuzzy weighted recurrence network
%       wPFRN: Prototype fuzzy weighted recurrence network
% Test:
%       x = randi([0 5],1,500); dim=3; tau=1; ncluster=3;
%       [wFRN,wPFRN]=fWrNet(x,dim,tau,ncluster);
%-------------------------------------------------------------


[nRow,~] = size(x);

if (nRow==1)
    x = x';
    [nRow,~] = size(x);
end

M=nRow-(dim-1)*tau;
PS=zeros(M,dim);

for i=1:M
    for j=1:dim
        PS(i,j)=x(i+(j-1)*tau);
    end
end

options = [2.0, 100, 1e-5, 0];
[center,FR,~] = fcm(PS, ncluster, options); % use FCM from Matlab

cRP=zeros(length(FR),length(FR),ncluster);
```

```
for c=1:ncluster
    for i=1:length(FR)
        for j=1:length(FR)
            if norm(PS(i,:) - PS(j,:)) == 0
                cRP(i,j,c)= 1; % reflexive
            elseif FR(c,i)>=FR(c,j)
                cRP(i,j,c)=FR(c,j);
            else
                cRP(i,j,c)=FR(c,i);
            end
        end
    end
end

FRP=zeros(length(cRP),length(cRP));
for c=1:ncluster
    for i=1:length(cRP)
        for j=1:length(cRP)
            if c==1
                FRP(i,j)=cRP(i,j,c);
            else
                if cRP(i,j,c)>=FRP(i,j)
                    FRP(i,j)=cRP(i,j,c);
                end
            end
        end
    end
end

%{
figure
imshow(imcomplement(FRP))
set(gca,'FontSize',14)
xlabel('Time')
ylabel('Time')
print('frp-si-', '-deps', '-r300');
%}

% FRP is fuzzy recurrence plot
% Fuzzy weighted recurrence network
I = eye(length(FRP)); % identity matrix
wFRN = FRP - I;

% Fuzzy weighted recurrence network of prototypes
Y=zeros(ncluster,ncluster,length(FR));

for c=1:length(FR)
    for i=1:ncluster
        for j=1:ncluster
            if norm(center(i,:) - center(j,:)) == 0
                Y(i,j,c)= 1; % reflexive
            elseif FR(i,c)>=FR(j,c)
                Y(i,j,c)=FR(j,c);
            else
                Y(i,j,c)=FR(i,c);
            end
        end
    end
end

B1 = zeros(ncluster,ncluster);

for c=1:length(FR)
```

```
    for i=1:ncluster
        for j=1:ncluster
            if c==1
                B1(i,j)=Y(i,j,c);
            else
                if Y(i,j,c)>=B1(i,j)
                    B1(i,j)=Y(i,j,c);
                end
            end
        end
    end
end

% B1 is fuzzy recurrence plot of prototypes
I = eye(length(B1)); % identity matrix
wPFRN = B1-I; % prototype fuzzy weighted recurrence network
```

## 5.2.2 *Illustrations*

These illustrations are based on the work reported in [4]. Two quantitative measures of a complex network are the average clustering coefficient and characteristic path length [5–7]. A clustering coefficient of a vertex (node) in a network expresses the magnitude that a node tends to cluster with other neighboring nodes. The average clustering coefficient expresses the average amount of connectivity around individual nodes of a network. The characteristic path length of a network is a measure of the efficiency of transfer of information in a network.

The average clustering coefficient for an unweighted network represented with an $N \times N$ (binary) adjacency matrix $\mathbf{A} = [a_{ij}]$, $i, j = 1, \dots, N$, is defined as

$$C = \frac{1}{N} \sum_{i=1}^{N} C_i, \tag{5.6}$$

where $C_i$ is the local unweighted clustering coefficient for node $i$, and defined as

$$C_i = \frac{\sum_{j,k} a_{ij} a_{jk} a_{ki}}{k_i(k_i - 1)}, \quad k_i \neq 0, 1, \tag{5.7}$$

where $k_i$ is the degree of node $i$, which is the number of links of node $i$.

The average clustering coefficient for a weighted network is defined as

$$C^w = \frac{1}{N} \sum_{i=1}^{N} C_i^w, \quad (5.8)$$

where $C_i^w$ is the local weighted clustering coefficient for node $i$, and defined as [8]

$$C_i^w = \frac{\sum_{j,k} [w_{ij} w_{ik} w_{jk}]^{1/3}}{k_i(k_i - 1)}, k_i \neq 0, 1, \quad (5.9)$$

where $w_{ij}$, $w_{ik}$, and $w_{jk} \in W$.

The characteristic path length of a network is defined as the average of all shortest path lengths:

$$L = \frac{1}{N(N-1)} \sum_{i \neq j, i, j=1}^{N} d_{ij}, \quad (5.10)$$

where $d_{ij}$ is the length of the shortest path between nodes $i$ and $j$, which can be computed using the Dijkstra's algorithm.

#### 5.2.2.1 The Lorenz System

Time series of 4000 data points of the $x$-, $y$-, and $z$-components of Lorenz system were used for the illustration of constructing FWRNs. Pink noise ($1/f$) and white Gaussian noise time series of the same length were also generated for comparing the performance of the recurrence network (RN) and FWRN. The embedding dimension and time delay chosen for both RN and FWRN were 3 and 1, respectively. The recurrence threshold for computing the RN was 10% of the mean of the time series. The partition entropy was used to determine an optimal number of clusters for the FCM algorithm. Based on the cluster validity, 22 clusters were selected as the number of nodes for constructing the FWRN.

The clustering coefficients and characteristic path lengths were computed for the RN and FWRN of the three Lorenz attractor components, pink noise, and white Gaussian noise. The results are shown in Table 5.1.

The clustering coefficients of the RN of the three Lorenz attractor components are higher than those of the two random time series. The characteristic path lengths of the RN of the $x$ and $y$ Lorenz attractor components are higher than those of the two random time series, whereas the $z$-component is lower than the two time series of noise. For the FWRN, the clustering coefficients and the characteristic path lengths of the three Lorenz attractor components are lower than those of the two random time series.

To visualize the relationship of the graph measures of the chaotic and random time series, phylogenetic trees of the two graph measures of the RN and FWRN were constructed using the hierarchical clustering with unweighted pair group method with arithmetic mean (UPGMA) [9], which are shown in Fig. 5.4. For the RN, the

**Table 5.1** Average clustering coefficients ($C$) and characteristic path lengths ($L$) of the Lorenz-system components and noise time series

| Time series | $C$ | $L$ | $C$ | $L$ |
|---|---|---|---|---|
| | RN | | FWRN | |
| $x$-component | 0.6845 | 13.0608 | 0.0152 | 0.0071 |
| $y$-component | 0.6890 | 14.0846 | 0.0226 | 0.0133 |
| $z$-component | 0.9313 | 1.1147 | 0.0255 | 0.0143 |
| Pink noise | 0.6618 | 2.1332 | 0.0440 | 0.0524 |
| White noise | 0.5538 | 2.3935 | 0.0483 | 0.0761 |

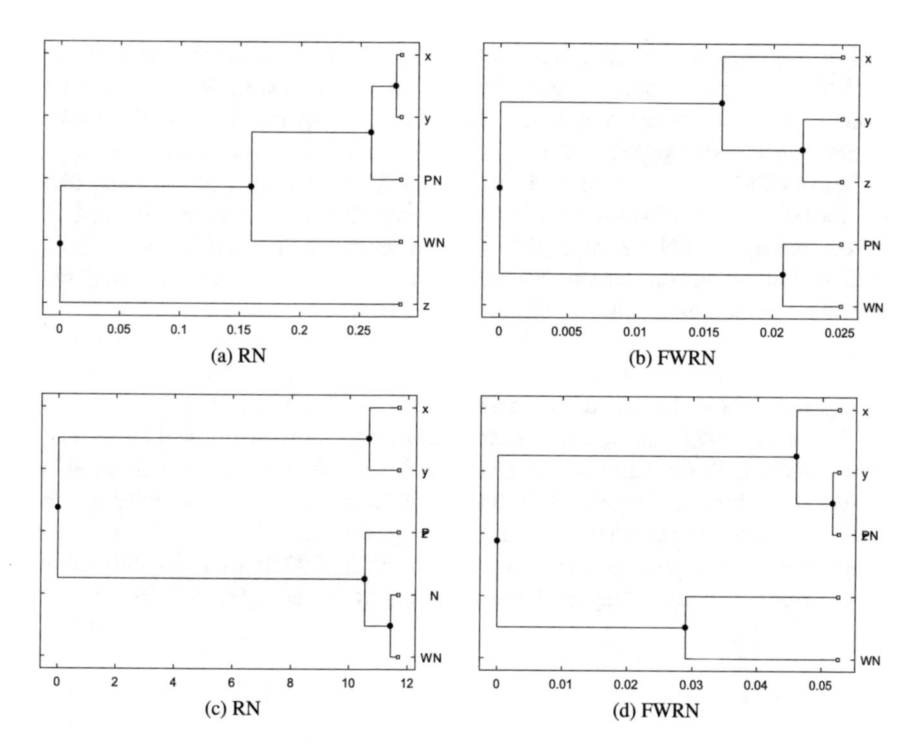


**Fig. 5.4** Hierarchical clustering of average clustering coefficients (**a**) and (**b**), and characteristic path lengths (**c**) and (**d**) for $x$, $y$, $z$ Lorenz components, pink noise (PN), and white Gaussian noise (WN) obtained from recurrence networks (RN) and fuzzy weighted recurrence networks (FWRN). (Figure reused from [4] with permission from the publisher)

clustering coefficients of the two random time series are located between the group of $x$- and $y$-components and the $z$-component (Fig. 5.4a), whereas for the FWRN, the tree well separates the three Lorenz attractor components from the two random time series (Fig. 5.4b). The tree locates the characteristic path lengths of the $z$-component in the same group with those of the two time series of noise obtained from the RN (Fig. 5.4c). For the FWRN, once again the tree show a clear separation between the three chaotic components and the two time series of noise (Fig. 5.4d) .

#### 5.2.2.2 EMG Signals

Another test was carried out to test and compare the performance of the proposed FWRN by using a publicly available PhysioBank dataset that includes electromyograms (EMG) recordings from three subjects: one is without neuromuscular disease (healthy), one with myopathy, and one with neuropathy described earlier. The length of the three EMG time series is 2000. The embedding dimension and time delay chosen for both RN and FWRN were 2 and 1, respectively. The recurrence threshold for computing the RN was also 10% of the mean of the time series. Based on the partition entropy for cluster validity, the number of clusters was selected as 35 that indicates the minimum value of the partition entropy.

The clustering coefficients and characteristic path lengths were computed for the RN and FWRN of the three EMG time series. These results are shown in Table 5.2.

Figure 5.5 shows the phylogenetic trees of the two graph measures of the RN and FWRN of the EMG time series recorded from healthy, myopathy, and neuropathy subjects. For both RN and FWRN, the clustering coefficients of the time series of the healthy subject are separated from those of the myopathy and neuropathy subjects (Fig. 5.5a, b). The characteristic path lengths of the RN for the healthy and neuropathy are grouped together, whereas those of the FWRN well split the healthy subject from the group of myopathy and neuropathy subjects (Fig. 5.5c, d).

**Table 5.2** Average clustering coefficients ($C$) and characteristic path lengths ($L$) of the EMG signals

| Subject | $C$ | $L$ | $C$ | $L$ |
|---|---|---|---|---|
| | RN | | FWRN | |
| Healthy | 0.8680 | 1.6977 | 0.0258 | 0.0334 |
| Myopathy | 1 | 1 | 0.0271 | 0.0370 |
| Neuropathy | 0.9442 | 1.3595 | 0.0272 | 0.0366 |

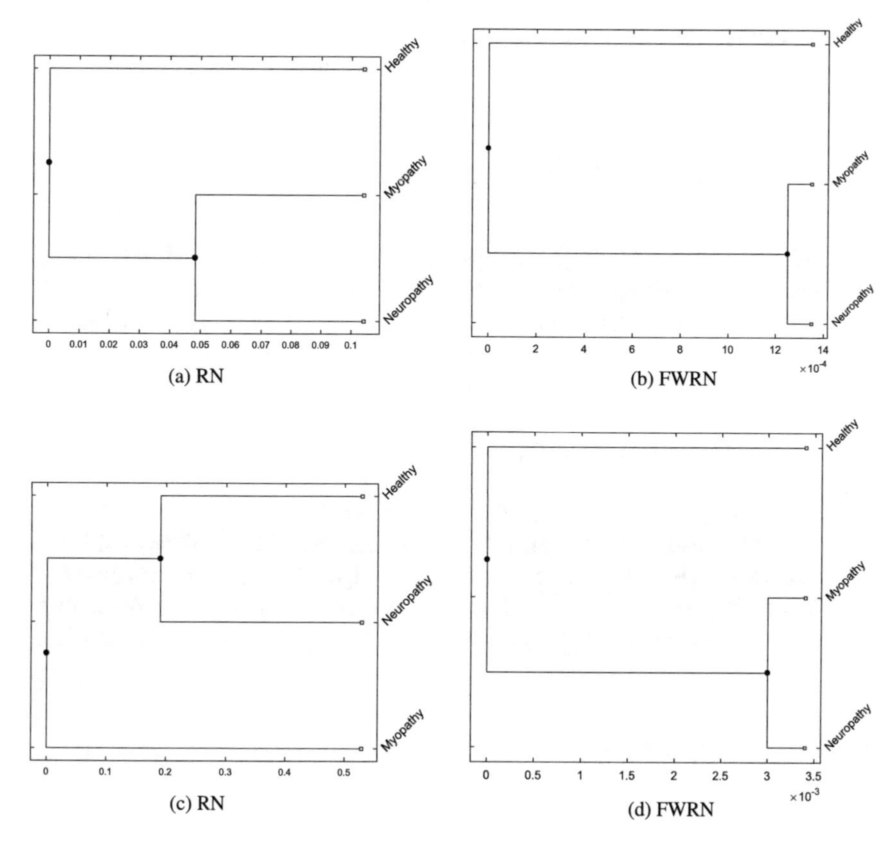


**Fig. 5.5** Hierarchical clustering of average clustering coefficients (**a**) and (**b**), and characteristic path lengths (**c**) and (**d**) for EMG recordings of heathy, myopathy, and neuropathy subjects obtained from recurrence networks (RN) and fuzzy weighted recurrence networks (FWRN). (Figure reused from [4] with permission from the publisher.)

## 5.3 Fuzzy Weighted Recurrence Networks of Multichannel Data

The term "channel" is a conventional expression used to refer to a certain component of an image. For example, an RGB image has three channels that are red (R), green (G) and blue (B) components. A grayscale image has only one channel. Let $\mathbf{I} = [f_{ijk}]$ be a multichannel image of size $M \times N \times K$, where $i = 1, \ldots, M$, $j = 1, \ldots, N$, and $k = 1, \ldots, K$. Let $m \geq 1$ be an integer, a local image window $\mathbf{W}^k_{ij} \in \mathbf{I}$ of size $(2m + 1) \times (2m + 1)$ is constructed for each pixel located at $ij$ in each of the $k$-components of the multichannel image, where $ij$ is the center of the window. This window can be considered as embedding dimensions in two-dimensional space, which considers the local spatial distribution around $f_{ij}$ of the $k$-

th image channel. The Frobenius norm can be used to transform each local window into a scalar measure that has the useful property of invariance under rotations as

$$\|\mathbf{W}_{ij}^k\|_F = \sqrt{\sum_{i-m}^{i+m}\sum_{j-m}^{j+m} |f_{ijk}|^2}, \tag{5.11}$$

where $(i-m), (j-m) > 0$, $(i+m) \leq M$, $(j+m) \leq N$, and any pixel at the center of the window that requires values from beyond the image boundaries is skipped.

We can then obtain a set of feature vectors $\mathbf{y}_{ij}$, $(i-m), (j-m) > 0$, by joining the Frobenius norms computed for each window of the $k$-th image channel at the same location, for example, a color image of three channels:

$$\mathbf{y}_{ij} = \left(\|\mathbf{W}_{ij}^1\|_F, \|\mathbf{W}_{ij}^2\|_F, \|\mathbf{W}_{ij}^3\|_F\right), \tag{5.12}$$

where $(i-m), (j-m) > 0$, $(i+m) \leq M$, $(j+m) \leq N$.

Since the Frobenius norm induced feature vector set $\mathbf{y}_{ij}$ can be computed for the multichannel image $\mathbf{I}$, the multichannel fuzzy weighted recurrence network (MC-FWRN) can be constructed using the same procedure for computing the FWRN, where each $\mathbf{y}_{ij}$ is considered as a state vector $\mathbf{x}_n$ of the phase-space reconstruction.

### *5.3.1 Computer Code*

**Matlab Program for Constructing a MC-FWRN**

```
function F = featureIm3D(I,m)

%{
Function to extract feature/state vectors from a color image.

Input: color image I, dimension m to create (2m+1)-by-(2m+1) window,
% where pixel of interest is at the window center.

Output: F is matrix of vectors of 3 color components of k pixels
of I, where each row of F is a segment used for quantifying image
irregularity\index{Irregularity}.

Note: Window is constructed for each pixel to account for local spatial
information around the pixel.

Author: Tuan Pham
Date: 21 July 2018
%}

% Split color image into 3 components
Im{1}=I(:,:,1);
Im{2}=I(:,:,2);
Im{3}=I(:,:,3);
```

```
[M,N] = size(Im{1});
W={};
F=[];

% Construct a window for each pixel and compute Frobenius norm
for t=1:3 % 3 color-image components
    G = Im{t};
    k=1;
    for i = m+1:M-m
        for j = m+1:N-m
            W{k} = G(i-m:i+m, j-m:j+m); % window with pixel k at center
            F(k,t) = norm(double(W{k}),'fro'); % Frobenius norm
            k=k+1;
        end
    end
end


function [acc,ad,G2] = fWrNetImage(PS,ncluster)
%-----------------------------------------------------------------------
% Fuzzy weighted recurrence networks of images (MC-FWRN)
% Reference: T.D. Pham
%-----------------------------------------------------------------------
% Input:
%       PS: preprocessed state vectors computed from "featureIm3D.m"
%       ncluster: number of clusters for fuzzy c-means clustering
%
% Output:
%
%       acc = average clustering coeff
%       ad = average shortest path length
%       G2 = graph of MC-FWRN
% Other Matlab Toolbox used: MatlabBGL by David Gleich
% https://mathworks.com/matlabcentral/fileexchange/10922-matlabbgl
%
%-------------------------------------------------------------


options = [2.0, 100, 1e-5, 0];

[center,FR,~] = fcm(PS, ncluster, options); % use fuzzy c-means from Matlab toolbox


% Fuzzy weighted recurrence network of prototypes
Y=zeros(ncluster,ncluster,length(FR));

for c=1:length(FR)
    for i=1:ncluster
        for j=1:ncluster
            if norm(center(i,:) - center(j,:)) == 0
                Y(i,j,c)= 1; % reflexive
            elseif FR(i,c)>=FR(j,c)
                Y(i,j,c)=FR(j,c);
            else
                Y(i,j,c)=FR(i,c);
            end
        end
    end
end

B1 = zeros(ncluster,ncluster);

for c=1:length(FR)
```

```
    for i=1:ncluster
        for j=1:ncluster
            if c==1
                B1(i,j)=Y(i,j,c);
            else
                if Y(i,j,c)>=B1(i,j)
                    B1(i,j)=Y(i,j,c);
                end
            end
        end
    end
end

% B1 fuzzy recurrence plot of prototypes
I = eye(length(B1)); % identity matrix
wPFRN = B1-I; % prototype fuzzy weighted recurrence network


% Compute statistics
G2 = graph(wPFRN);

% Use sparse function to construct weighted adjacency matrix
% representation of the graph.
nn2 = numnodes(G2);
[s2,t2] = findedge(G2);
A2 = sparse(s2,t2,G2.Edges.Weight,nn2,nn2);

%{
figure
plot(G2,'EdgeLabel',G2.Edges.Weight) % plot graph prototype only
%}


% Average shortest path
D2=distances(G2); % Dijkstra algorithm
N2=length(wPFRN);
ad = sum(sum(D2))/(N2*(N2-1));


% Average clustering coeff using "MatlabBGL" toolbox
cc2 = clustering_coefficients(A2);
acc = mean(cc2);
```

---

### 5.3.2 *Illustrations*

The recovery from a burn injury depends on the level of severity, where more severe burns require the emergency of medical treatment in order to avoid complications and death. After burn healing, scarring is a concerning issue of both body image and physical function to the patients. In addition, the quantification of the characteristics of burn scars has important implications for monitoring of the healing process, comparison and assessment of different surgical interventions that provide useful information for more optimal treatment and effective preoperative counsel-

**Table 5.3** The Vancouver Scar Scale (VSS)

| Scar characteristic | | Score |
|---|---|---|
| Vascularity | Normal | 0 |
| | Pink | 1 |
| | Red | 2 |
| | Purple | 3 |
| Pigmentation | Normal color | 0 |
| | Hypopigmentation | 1 |
| | Hyperpigmentation | 2 |
| Pliability | Normal | 0 |
| | Supple | 1 |
| | Yielding | 2 |
| | Firm | 3 |
| | Ropes | 4 |
| | Contracture | 5 |
| Height | Flat | 0 |
| | <2 mm | 1 |
| | 2–5 mm | 2 |
| | >5 mm | 3 |
| Total score | | 13 |

ing. Therefore, it is important to evaluate the severity of burn scars, and several tools and instruments have been developed for assessing one or more aspects of burn scars to improve the quality of life of the patient [10].

For burn-scar treatment, it is critical to make accurate and reproducible clinical findings of scar assessment so that various interventions and treatments can be consistently interpreted and compared on a universal basis [11]. Since the introduction of the Vancouver Scar Scale (VSS) [12], more than ten scar rating scales have been developed to contribute to the standardization of scar therapy and enhancement of the assessment [11]. Table 5.3 shows the scar characteristics and their VSS-based scores, based on which scarring after burn healing can be assessed.

A review of scar scales and scar measurement instruments suggested that there is a need for the development of an optimal scar scoring system so that pathologic scarring can be better treated [13]. In addition to the subjective assessment of burn scars, objective scar evaluations have been carried out by measuring the physical properties of the scar such as its height or its vascularity. General criteria for assessing scar severity are based on color, dimensions, texture, biomechanical properties, patho-

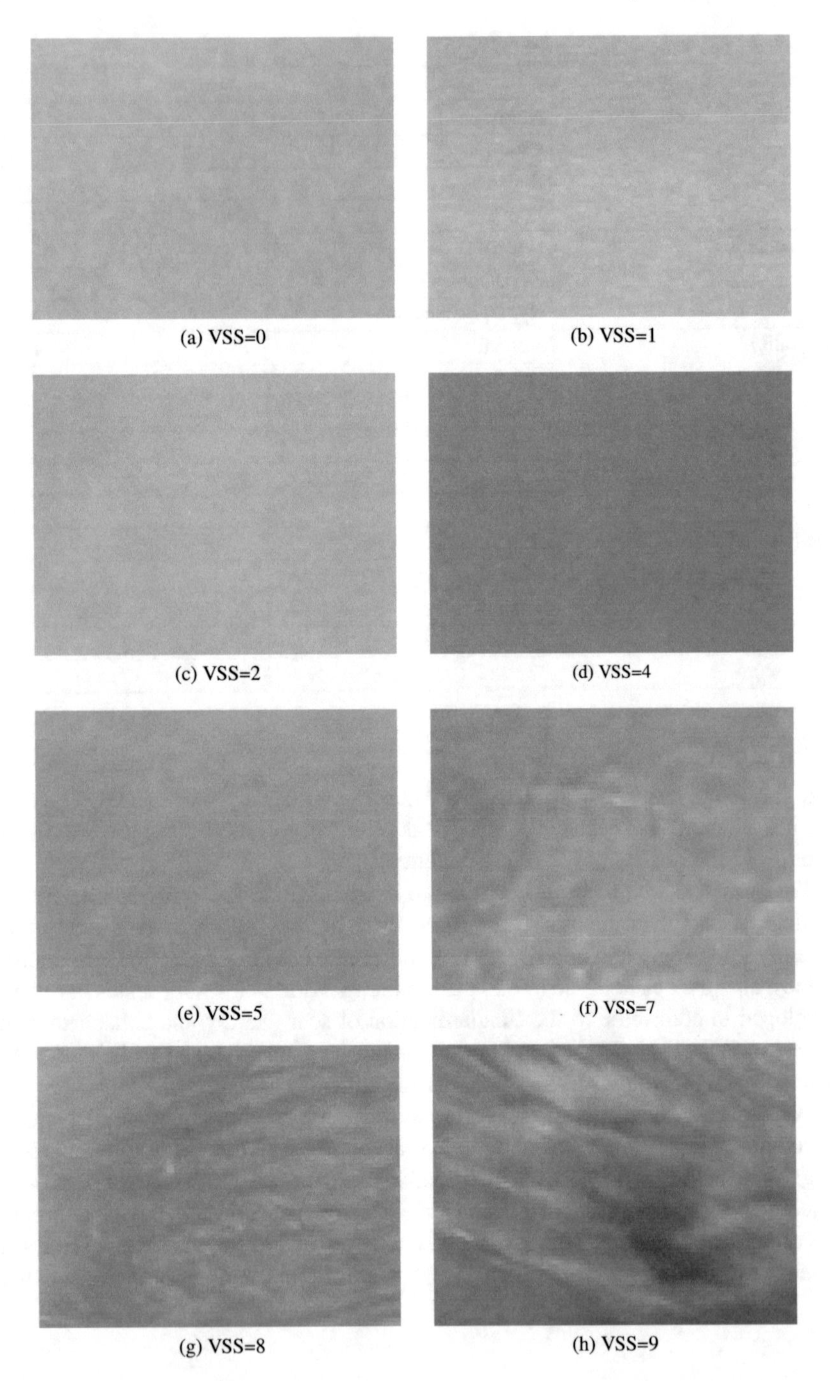


**Fig. 5.6** Burn-scar images

physiological disturbances (oxygen tension, water loss, and moisture content), tissue microstructure, and pain [14]. However, it is still difficult to obtain an overall scar rating using objective measurement devices because many expensive instruments such as the tonometer, dermaspectrometer, or chromometer are required, in addition to the requirement of experienced users, making the assessment time-consuming and impractical in busy clinical settings [15–17].

A pioneering attempt for developing an automated burn-scar assessment approach was reported in [18] that extracts texture and color features of burn-scar images for different VSS-based scores and uses the error-correcting output coding in machine learning for assigning the scar images to corresponding total VSS-based scores. An accuracy of 85% was obtained, while 92 and 98% were achieved for the tolerances of one VSS-based score and two VSS-based scores, respectively.

Figure 5.6 shows small samples of burn scar and their VSS-based scores used in [18]. Each RGB image is of size $69 \times 92 \times 3$. As an illustration of constructing MC-FWRNs of these color images, the dimension $m = 3$ to create $(2m + 1)$-by-$(2m + 1)$ window, where the pixel of interest is at the window center, and the number of clusters = 10, which is the number of nodes of each MC-FWRN. The average clustering coefficients ($C$) and characteristic path lengths ($L$) computed from the MC-FWRNs of the burn-scar images are shown in Table 5.4. Figure 5.7 shows the MC-FWRNs of the burn-scar images. The average clustering coefficients and characteristic path lengths can be used as compact features for machine learning and classification of burn scar images. Since by definition, MC-FWRNs are fully connected networks, the MC-FWRNs of the burn-scar images can be further constructed using a cutoff weight that sets zero to edge weights being less than the cutoff value and may further differentiate the networks of different VSS-based scores in terms of the network topology.

**Table 5.4** Average clustering coefficients ($C$) and characteristic path lengths ($L$) of burn-scar images

| VSS-based score | $C$ | $L$ |
|---|---|---|
| 0 | 0.0647 | 0.0687 |
| 1 | 0.0589 | 0.0560 |
| 2 | 0.0804 | 0.1209 |
| 4 | 0.0584 | 0.0550 |
| 5 | 0.0574 | 0.0537 |
| 7 | 0.0607 | 0.0619 |
| 8 | 0.0747 | 0.0990 |
| 9 | 0.0538 | 0.0412 |

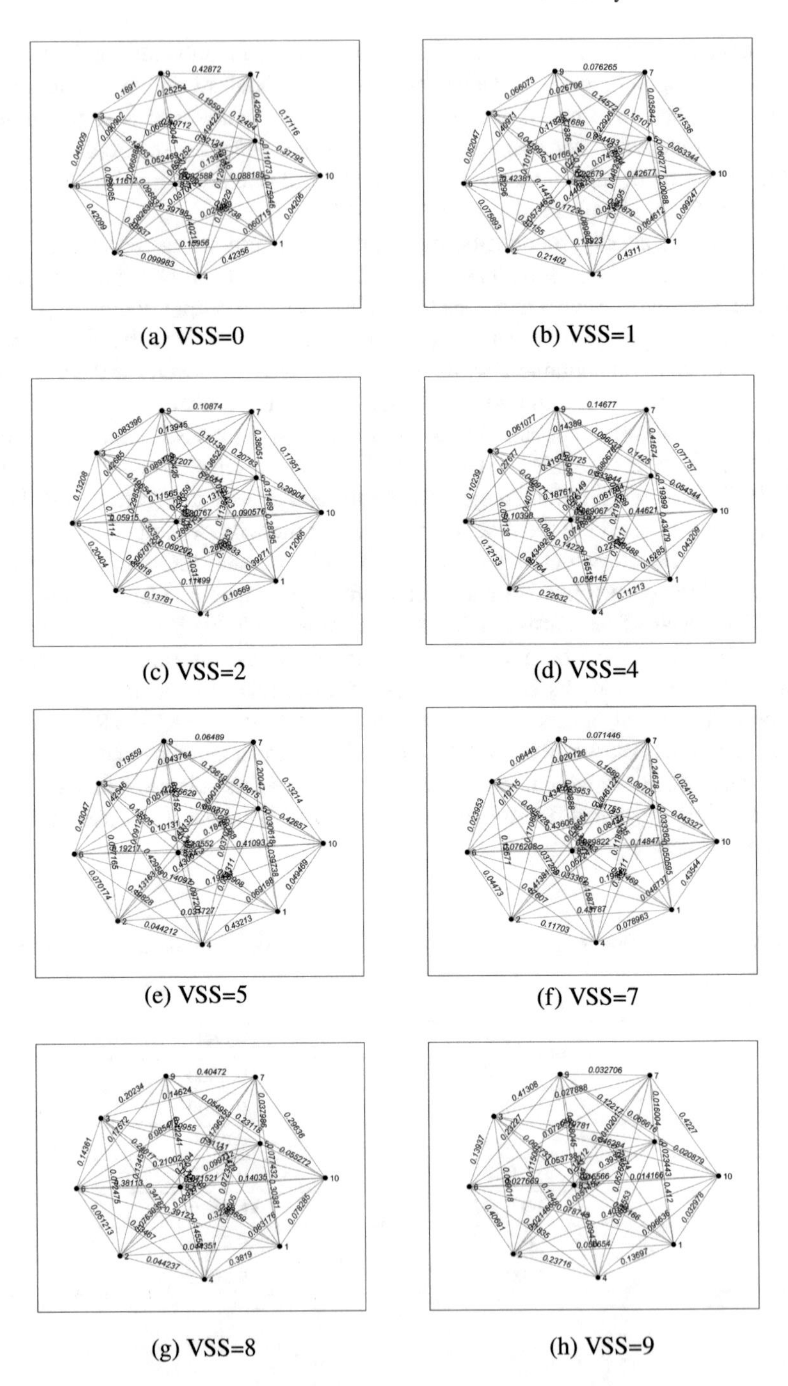

(a) VSS=0

(b) VSS=1

(c) VSS=2

(d) VSS=4

(e) VSS=5

(f) VSS=7

(g) VSS=8

(h) VSS=9

**Fig. 5.7** MC-FWRNs of burn-scar images

## References

1. Marwan N et al (2009) Complex network approach for recurrence analysis of time series. Phys Lett A 373:4246–4254
2. Pham TD (2017) From fuzzy recurrence plots to scalable recurrence networks of time series. EPL (Europhys Lett) 118:20003
3. Examples of electromyograms. PhysioNet. https://physionet.org/physiobank/database/emgdb. Accessed 6 Feb 2018
4. Pham TD (2019) Fuzzy weighted recurrence networks of time series. Phys A 513:409–417
5. Watts DJ, Strogatz S (1998) Collective dynamics of "small-world" networks. Nature 393:440–442
6. Barrat A et al (2004) The architecture of complex weighted networks. Proc Natl Acad Sci 101:3747–3752
7. Albert R, Barabasi AL (2002) Statistical mechanics of complex networks. Rev Mod Phys 74:47–97
8. Fagiolo G (2007) Clustering in complex directed networks. Phys Rev E 76:026107
9. Weib M, Goker M (2011) Molecular phylogenetic reconstruction. In: Kurtzman CP, Fell JW, Boekhout T (eds) The yeasts, 5th edn. Elsevier, London
10. Brusselaers N et al (2010) Burn scar assessment: a systematic review of objective scar assessment tools. Burns 36:1157–1164
11. Nguyen TA et al (2015) A review of scar assessment scales. Semin Cutan Med Surg 34:28–36
12. Sullivan T et al (1990) Rating the burn scar. Burn Care Rehabil 11:256–260
13. Fearmonti R et al (2010) A review of scar scales and scar measuring devices. Eplasty 10:e43
14. Lee KC et al (2016) A systematic review of objective burn scar measurements. Burns Trauma 4:14
15. Tyack Z et al (2012) A systematic review of the quality of burn scar rating scales for clinical and research use. Burns 38:6–18
16. Oliveira GV et al (2005) Objective assessment of burn scar vascularity, erythema, pliability, thickness, and planimetry. Dermatol Surg 31:48–58
17. Chae JK et al (2016) Values of a patient and observer scar assessment scale to evaluate the facial skin graft scar. Ann Dermatol 28:615–623
18. Pham TD et al (2017) Automated VSS-based burn scar assessment using combined texture and color features of digital images in error-correcting output coding. Sci Rep 7:16744

# Chapter 6
# Entropy Algorithms

## 6.1 Approximate Entropy and Sample Entropy

The concept of the approximate entropy (ApEn) [2] was developed for quantifying irregularity or predictability in time series. Consider a time series $X$ of length $N$ taken at regular intervals: $X = (x_1, x_2, \dots, x_N)$, and a given embedding dimension $m$, a set of newly reconstructed time series from $X$, denoted as $Y$, can be established as $Y = (y_1, y_2, \dots, y_{N-m+1})$, where $y_i = (x_i, x_{i+1}, \dots, x_{i+m-1})$, $i = 1, 2, \dots, N - m + 1$. Given a positive tolerance value $r$, the probability of vector $y_i$ being similar to vector $y_j$ is computed as

$$C_i^m(r) = \frac{1}{N-m+1} \sum_{j=1}^{N-m+1} \theta[d(y_i, y_j)], \tag{6.1}$$

where $\theta(d(y_i, y_j))$ is the step function defined as

$$\theta[d(y_i, y_j)] = \begin{cases} 1 : d(y_i, y_j) \le r \\ 0 : d(y_i, y_j) > r \end{cases} \tag{6.2}$$

The distance between the two vectors can be obtained by

$$d(y_i, y_j) = \max_k(|x_{i+k-1} - x_{j+k-1}|), k = 1, 2, \dots, m. \tag{6.3}$$

The probabilities of all vectors that are similar to one another are computed as

$$C^m(r) = \frac{1}{N-m+1} \sum_{i=1}^{N-m+1} \log[C_i^m(r)]. \tag{6.4}$$



T. D. Pham, *Fuzzy Recurrence Plots and Networks with Applications in Biomedicine*, https://doi.org/10.1007/978-3-030-37530-0_6

This leads to the definition of ApEn expressed as

$$\text{ApEn} = C^m(r) - C^{m+1}(r). \tag{6.5}$$

For the calculation of ApEn, $m = 2$ or $3$ and $r = 0.1$ to $0.25 \times \sigma$, where $\sigma$ is the standard deviation of the time series, are suggested as appropriate for the implementation of ApEn [3].

Because ApEn includes self-matching, its computation can result in bias. A modified version of ApEn known as the sample entropy (SampEn) [4] was introduced to remove self-matching.

Let $B_i^m(r)$ be defined as

$$B_i^m(r) = \frac{1}{N-m-1} \sum_{j=1}^{N-m} \theta[d(y_i, y_j)], i \neq j. \tag{6.6}$$

Thus, $B^m(r)$ is given by

$$B^m(r) = \frac{1}{N-m} \sum_{i=1}^{N-m} B_i^m(r). \tag{6.7}$$

The formulation of SampEn is expressed as

$$\text{SampEn} = -\log\left[\frac{B^{m+1}(r)}{B^m(r)}\right]. \tag{6.8}$$

## 6.2 Multiscale Entropy

The multiscale entropy (MSE) [5] was developed to measure entropy such as SampEn at different scales by averaging nonoverlapping time points of the original time series in order to better reveal patterns of predictability or regularity in the time series. The purpose of MSE analysis of time series is to overcome the inconsistency encountered with single-scale analysis in increasing and decreasing entropy values of certain physiological signals. The MSE works by applying a "coarse-graining" process to the original time series to generate several time series of different scales, and then computing the SampEn for all coarse-grained time series, which can be plotted as a function of the scale factor.

Consider the time series $X$ of length $N$: $X = (x_1, x_2, \ldots, x_N)$. For a scale factor $\tau$, a new time series $X^\tau$ is created by the MSE as follows [5]:

$$X_j^\tau = \frac{1}{\tau} \sum_{i=(j-1)\tau+1}^{j\tau} x_i, \; 1 \leq j \leq N/\tau. \tag{6.9}$$

For each time series $X_j^\tau$, a SampEn value can be computed accordingly. Then a sequence of SampEn values obtained from different scale factors constitutes the profile of MSE.

## 6.3 Time-Shift Multiscale Entropy

The time-shift multiscale entropy (TSME) [1] was motivated by the Higuchi's fractal dimension (HFD) [6] for computing the fractal dimension of irregular time series. The HFD computes the "mean length" of the curve of a time series by constructing a set of new time series that has the property of a fractal curve over all time scales as each time series can be considered a reduced scale form of the whole.

A set of new time series constructed from the original time series by the HFD are considered as the phase distribution. This phase distribution can reveal strong effects of the irregularity of time series [7]. It has been reported that the HFD is a stable numerical approach for time series analysis, including stationary, nonstationary, deterministic, and stochastic signals [8–11]. Therefore, the time-shift entropy can be expected to be applied to these types of signals for studying irregularity in time series.

Once again consider a time series $X$ of length $N$: $X = (x_1, x_2, \ldots, x_N)$, and let $\beta$ and $k$ be positive integers, where $\beta = 1, 2, \ldots, k$, then $k$ new time series can be generated using the following [6]:

$$X_k^\beta = (x_\beta, x_{\beta+k}, x_{\beta+2k}, \ldots, x_{\beta+\lfloor \frac{N-\beta}{k} \rfloor k}), \tag{6.10}$$

where $\lfloor \frac{N-\beta}{k} \rfloor$ is the "floor" function that rounds $\frac{N-\beta}{k}$ to the largest integer not exceeding $\frac{N-\beta}{k}$.

From Eq. (6.10), $\beta$ and $k$ indicate the initial time point and time interval, respectively, that is, for a given time interval $k$, $k$ new time series are constructed using $k$ time shifts. For example, [12], let $k = 3$ and $N = 100$, three new time series are generated as follows:

$$X_{k=3}^{\beta=1} : (x_1, x_4, x_7, \ldots, x_{100}),$$
$$X_{k=3}^{\beta=2} : (x_2, x_5, x_8, \ldots, x_{98}),$$
$$X_{k=3}^{\beta=3} : (x_3, x_6, x_9, \ldots, x_{99}).$$

The TSME works by constructing $k$ time-shift series for a given time interval $k$, then computing either SampEn or ApEn for all time-shift time series, denoted as $\text{TSME}_k^\beta$, $\beta = 1, \ldots, k$. The TSME for each $k$, denoted as $\text{TSME}_k$, $k = 1, \ldots, k_{\max}$, is defined as the average value of all $\text{TSME}_k^\beta$, that is,

$$\text{TSME}_k = \frac{1}{k}\sum_{\beta=1}^{k} \text{TSME}_k^{\beta}. \tag{6.11}$$

The procedure for computing TSME is described as follows:

1. Given a time series $X$, dimension $m$, tolerance $r$, and $k_{\max}$.
2. Set $k = 1$.
3. Using Eq. (6.10) to construct $k$ time-shift time series from $X$: $X_k^{\beta}$, $\beta = 1, \ldots, k$.
4. For each $X_k^{\beta}$, $\beta = 1, \ldots, k$, compute $\text{TSME}_k^{\beta}$.
5. Compute $\text{TSME}_k$ using Eq. (6.11).
6. Set $k = k + 1$.
7. Repeat steps 3–6 until $k = k_{\max}$.

## 6.4 Illustrations

### 6.4.1 *Analysis of Time Series with Known Properties*

In signal processing, white noise and $1/f$ (pink) noise are signals with known statistical properties, which are widely found as good approximation of many real-world data and generate mathematically tractable models. Time series of white noise are sequences of serially uncorrelated random variables with zero mean and finite variance, having equal intensity at different frequencies and constant power spectral density [13]. Time series of $1/f$ noise is a signal with a frequency spectrum such that the power spectral density is inversely proportional to the frequency of the signal; it is an intermediate between the white noise and random walk noise with no correlation between increments [14].

The data include the signals of white noise (mean = 0 and variance = 1) and $1/f$ noise, both consisting of 10,000 time steps. Another type of signal of known property used in this study is the Lorenz attractor [15] that consists of $x$ (convection velocity), $y$ (temperature difference), and $z$ (temperature gradient) components. These three signals, each has the length of 4000 time steps, behave irregularly with time: the $x$- and $y$-components fluctuate around positive and negative values, and the $z$-component oscillates around the range from about 10 to 40. Setting $m = 2$ and $r = 0.15 \times \sigma$, where $\sigma$ is the standard deviation of the original time series, to compute ApEn and SampEn.

Tables 6.1 and 6.2 show the $\text{TSME}_k^{\beta}$ values of the white noise using ApEn and SampEn, respectively. With $k_{\max} = 10$, the ApEn-based $\text{TSME}_k^{\beta}$ tends to slightly decrease with increasing $k$ from 2.4 for $k = 1$ to 1.4 for $k = 10$, while the SampEn-based $\text{TSME}_k^{\beta}$ keeps fairly constant around the value of 2.4. Figure 6.1 shows the $\text{TSME}_k$ of the white noise and $1/f$ noise, respectively, where the SampEn-based $\text{TSME}_k$ for both signals are fairly constant. Both SampEn-based $\text{TSME}_k^{\beta}$ and SampEn-based $\text{TSME}_k$ meet the expectation for the analysis of randomly generated time series. The

**Table 6.1** ApEn-based $TSME_k^\beta$ of white noise

| k | β | | | | | | | | | |
|---|---|---|---|---|---|---|---|---|---|---|
| | 1 | 2 | 3 | 4 | 5 | 6 | 7 | 8 | 9 | 10 |
| 1 | 2.3566 | | | | | | | | | |
| 2 | 2.1859 | 2.1829 | | | | | | | | |
| 3 | 2.0636 | 2.0612 | 2.0634 | | | | | | | |
| 4 | 1.9363 | 1.9181 | 1.9475 | 1.9482 | | | | | | |
| 5 | 1.8376 | 1.8341 | 1.8554 | 1.8366 | 1.8234 | | | | | |
| 6 | 1.7671 | 1.7374 | 1.7496 | 1.7648 | 1.7877 | 1.7441 | | | | |
| 7 | 1.6823 | 1.6779 | 1.7000 | 1.6611 | 1.7120 | 1.6998 | 1.6505 | | | |
| 8 | 1.6358 | 1.6251 | 1.6094 | 1.6088 | 1.6103 | 1.6208 | 1.6151 | 1.6390 | | |
| 9 | 1.5812 | 1.5524 | 1.5472 | 1.5793 | 1.5582 | 1.5655 | 1.5456 | 1.5649 | 1.5294 | |
| 10 | 1.4817 | 1.4770 | 1.4810 | 1.4789 | 1.4744 | 1.4782 | 1.5036 | 1.4848 | 1.4731 | 1.5091 |

**Table 6.2** SampEn-based $TSME_k^\beta$ of white noise

| k | β | | | | | | | | | |
|---|---|---|---|---|---|---|---|---|---|---|
| | 1 | 2 | 3 | 4 | 5 | 6 | 7 | 8 | 9 | 10 |
| 1 | 2.4711 | | | | | | | | | |
| 2 | 2.4781 | 2.4752 | | | | | | | | |
| 3 | 2.4900 | 2.4937 | 2.4601 | | | | | | | |
| 4 | 2.4412 | 2.4113 | 2.4768 | 2.4832 | | | | | | |
| 5 | 2.4577 | 2.4458 | 2.4906 | 2.4608 | 2.4204 | | | | | |
| 6 | 2.4764 | 2.4421 | 2.4448 | 2.5061 | 2.5306 | 2.4517 | | | | |
| 7 | 2.4714 | 2.4664 | 2.5163 | 2.3949 | 2.5100 | 2.4686 | 2.4581 | | | |
| 8 | 2.5057 | 2.4644 | 2.4664 | 2.4603 | 2.4300 | 2.4542 | 2.4682 | 2.5132 | | |
| 9 | 2.5667 | 2.4541 | 2.4202 | 2.5461 | 2.4505 | 2.4945 | 2.3994 | 2.4986 | 2.4064 | |
| 10 | 2.4300 | 2.3704 | 2.4166 | 2.3865 | 2.3618 | 2.3223 | 2.4288 | 2.4217 | 2.3468 | 2.5132 |

trend in ApEn-based $TSME_k^\beta$ and ApEn-based $TSME_k$ should be due to the bias of self-matching in the computation of ApEn [4].

Figure 6.2 shows that $TSME_k^\beta$ using either ApEn or SampEn can distinguish white noise and $1/f$ noise from the chaotic time series of the Lorenz system. Similarly, Fig. 6.3 shows that $TSME_k$ using either ApEn or SampEn can distinguish white noise and $1/f$ noise from the chaotic time series. Particularly, $TSME_k$ can also separate white noise from $1/f$ noise. $TSME_k^\beta$ and $TSME_k$ values of white noise and $1/f$ noise are higher than those of the three chaotic time series as shown in Figs. 6.2 and 6.3, respectively, which suggest the reliability of the TSME analysis as chaotic signals are deterministic and therefore more predictable than noise data.

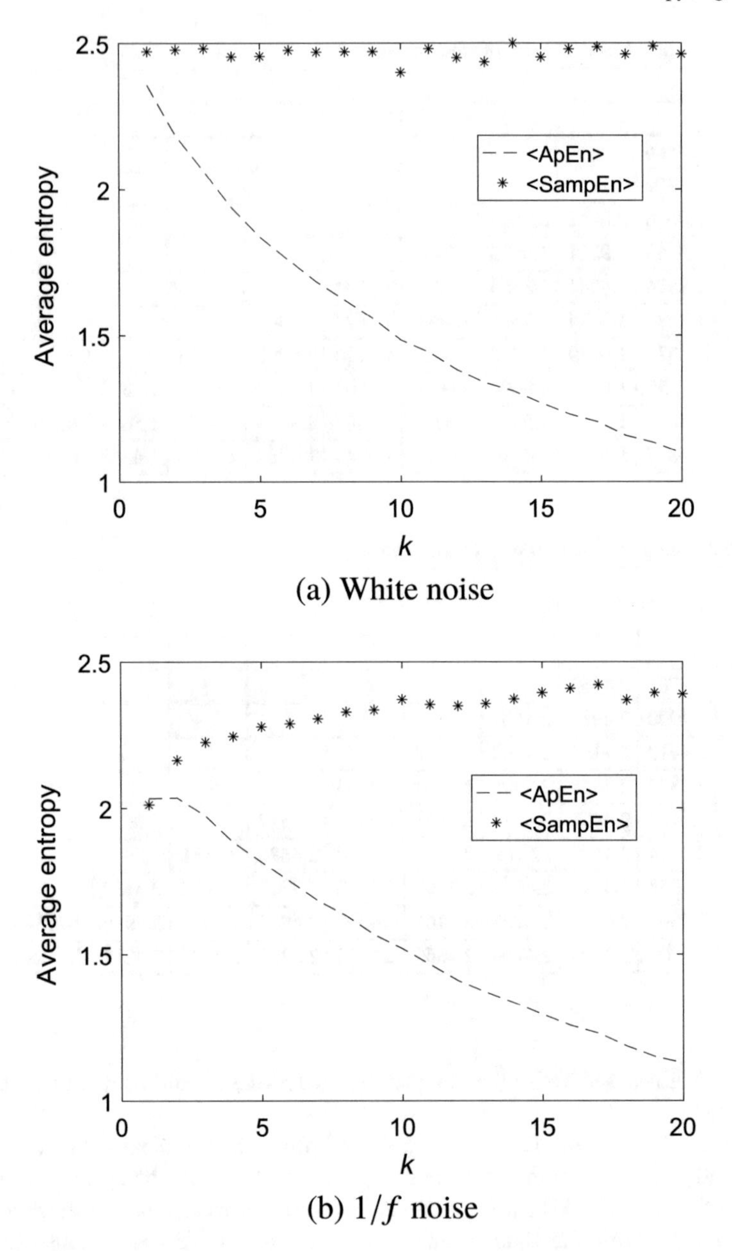


**Fig. 6.1** $\mathrm{TSME}_k$ obtained from noise time series, where $< \cdot >$ stands for average

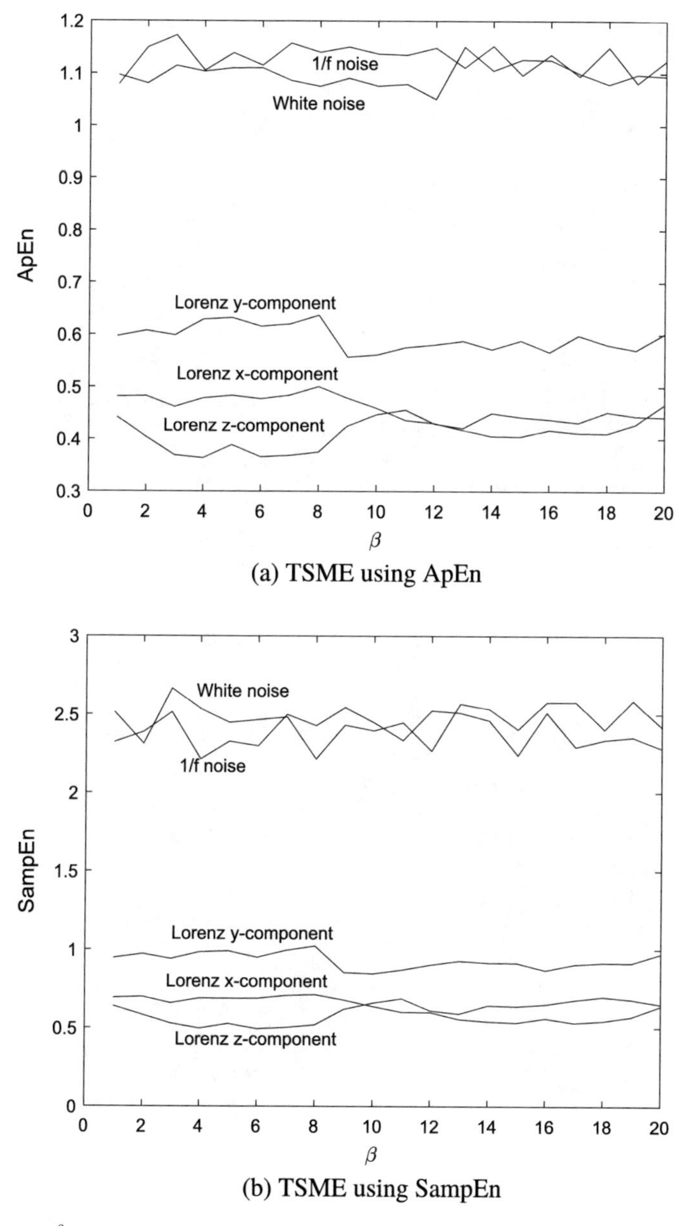


(a) TSME using ApEn

(b) TSME using SampEn

**Fig. 6.2** $\text{TSME}_{k=20}^{\beta}$ of white noise, $1/f$ noise, and the Lorenz system

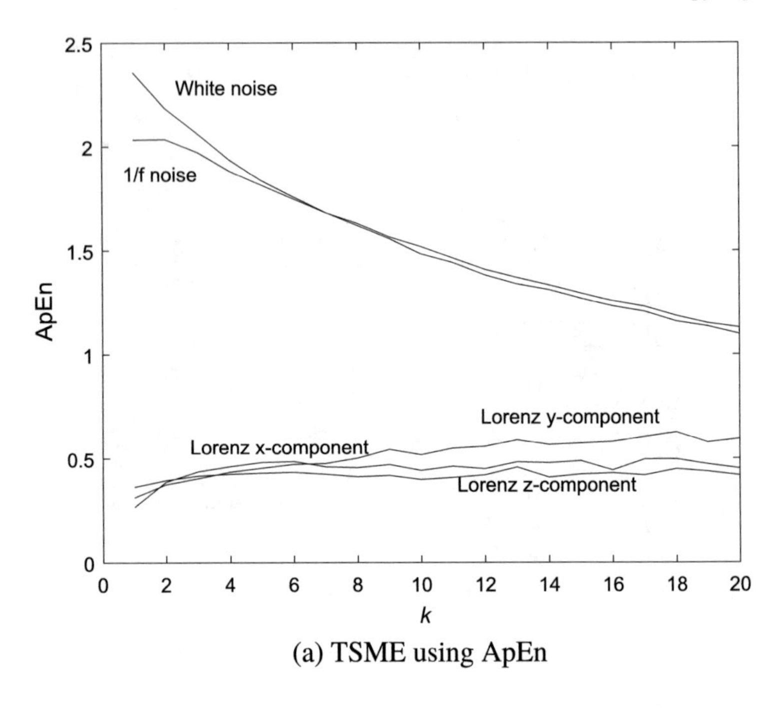


(a) TSME using ApEn

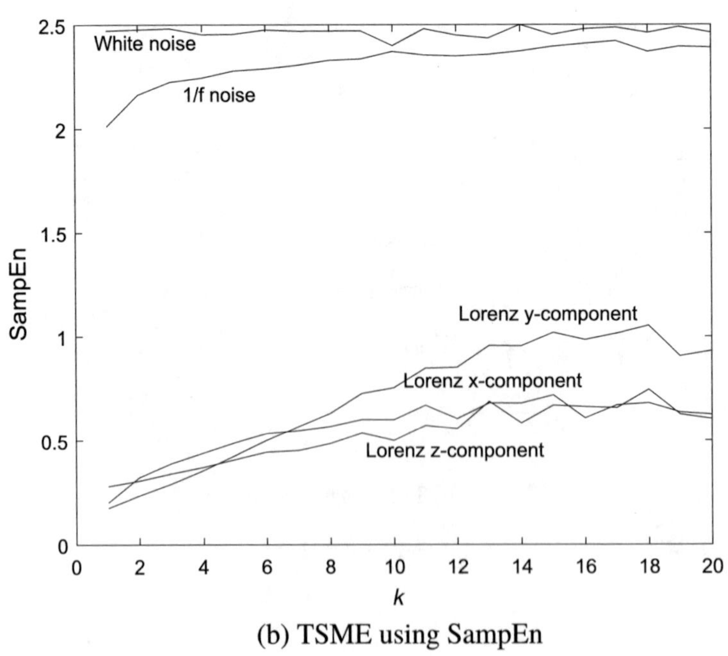


(b) TSME using SampEn

**Fig. 6.3** TSME$_k$ ($k_{\max} = 20$) of white noise, $1/f$ noise, and the Lorenz system

### 6.4.2 Analysis of Physiological Signals

Other computer experiments were carried out using the photoplethysmography (PPG) and electromyography (EMG) signals. The pulses of the index fingers of the left hands of 43 elderly participants and the middle-aged caregiver were synchronously measured with a PPG sensor and studied in [16] for automated assessment of therapeutic communication for cognitive stimulation for people with cognitive decline. The EMG signals were obtained from the Physical Action Data Set [17], where the channel measured on the right bicep of the participants with the Delsys EMG wireless apparatus was used in this study, for the classification of normal and aggressive human physical actions.

Figure 6.4 shows the first 5000 samples of synchronized PPG signals of two elderly participants and the caregiver. Figure 6.5 shows the first 5000 samples of the EMG signals of six normal actions: bowing, clapping, handshaking, hugging, jumping, and running. Figure 6.6 shows the first 5000 samples of the EMG signals of six aggressive actions: elbowing, front kicking, hammering, kneeing, pulling, and punching.

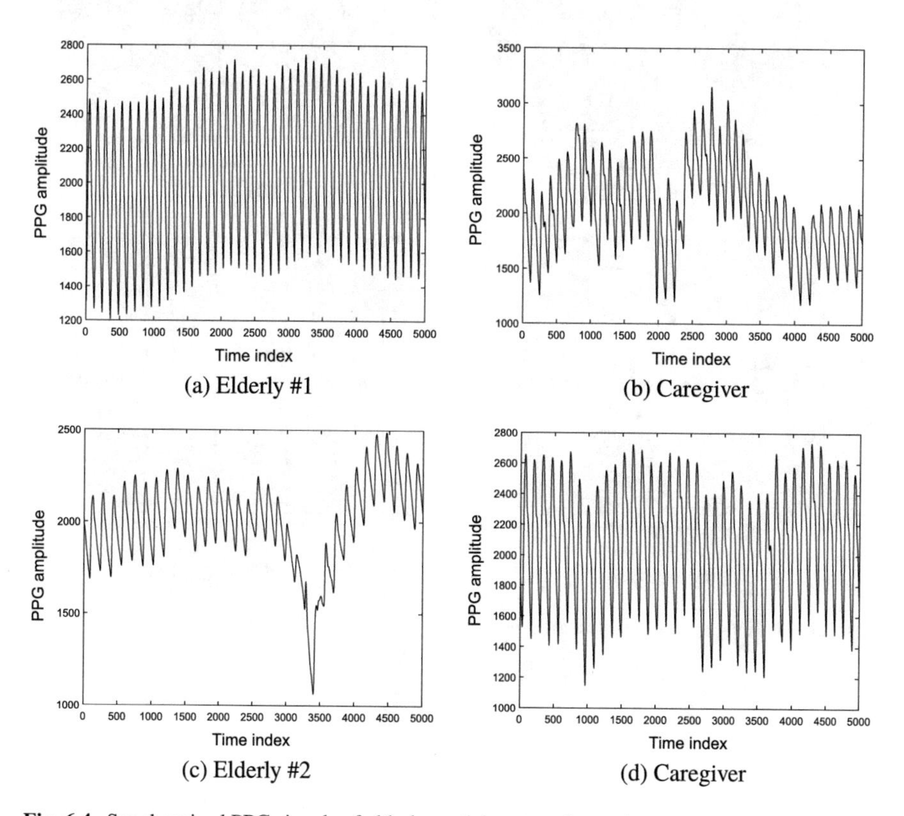


**Fig. 6.4** Synchronized PPG signals of elderly participants and caregiver

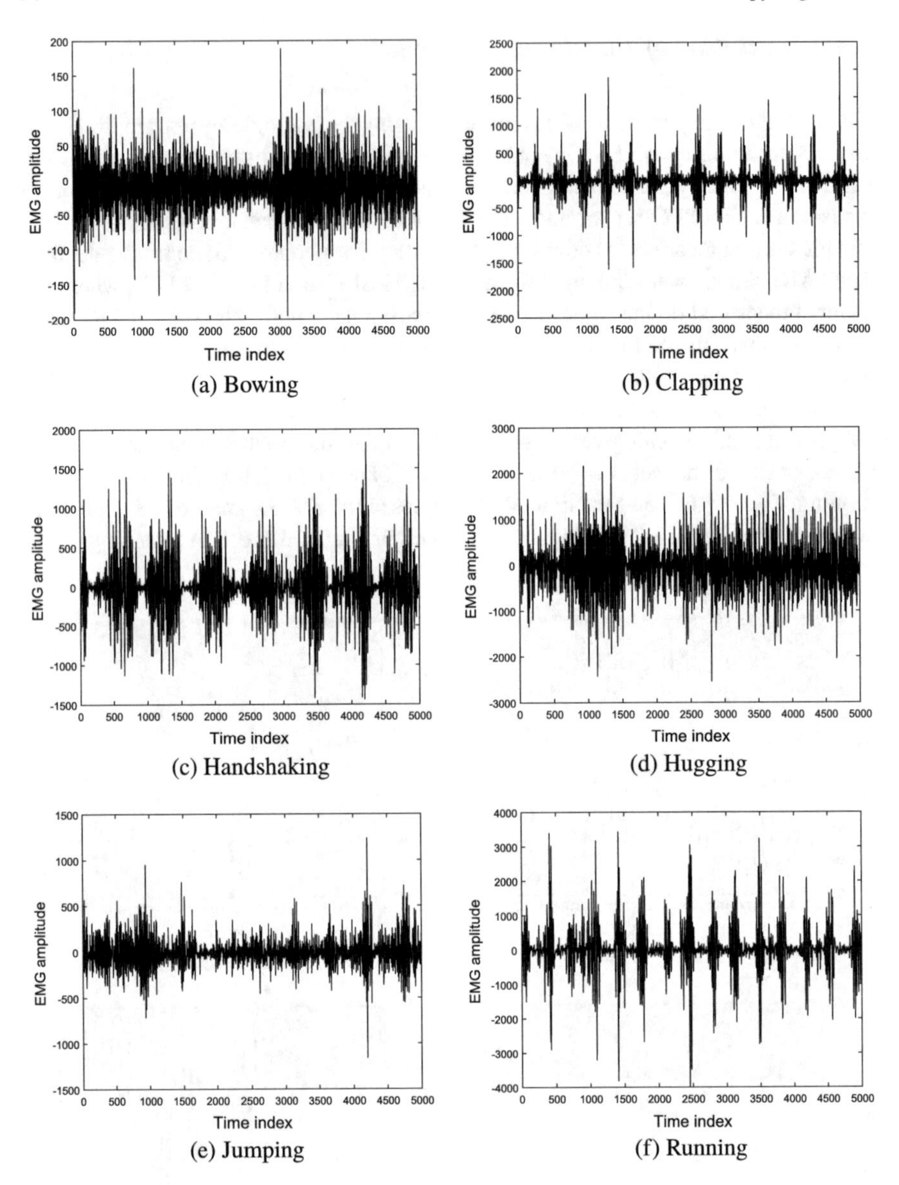


**Fig. 6.5** EMG signals of normal human actions

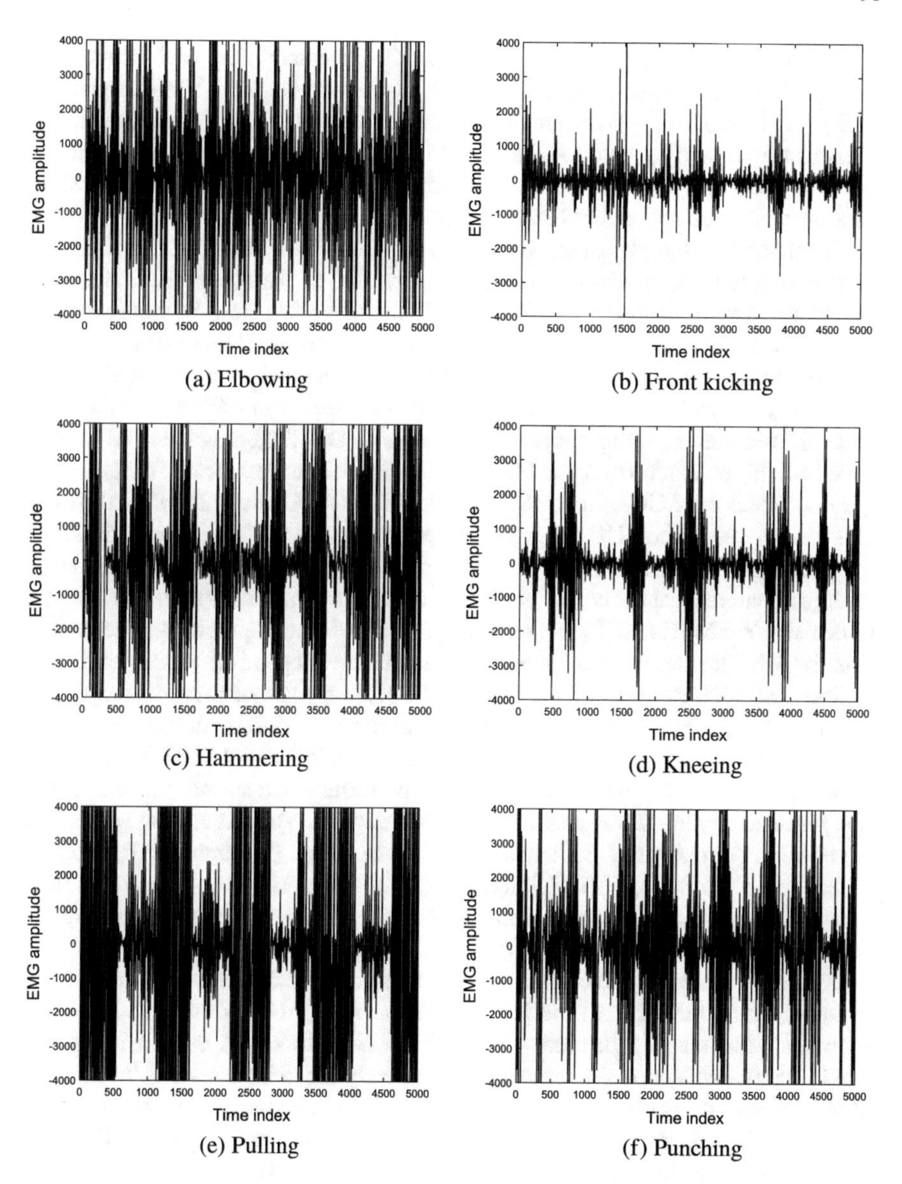


(a) Elbowing

(b) Front kicking

(c) Hammering

(d) Kneeing

(e) Pulling

(f) Punching

**Fig. 6.6** EMG signals of aggressive human actions

For the MSE analysis, $\tau = 20$ to compute $X_j^\tau$ expressed in Eq. (6.9) as the MSE feature of 20 scale factors. For the TSME, $k_{\max} = 20$ to compute $\text{TSME}_k$ expression in Eq. (6.11) as the TSME feature of 20 phase shifts. Only SampEn was used in this analysis for computing both MSE and TSME, with $m = 2$ and $r = 0.15 \times \sigma$, where again $\sigma$ is the standard deviation of the original time series. Both PPG and EMG datasets were used for pattern classification.

The first 5,000 samples of the synchronized, detrended PPG signals of 43 elderly participants and the middle-aged caregiver were used to classify the PPG signals of the elderly from the caregiver. MSE and TSME features using SampEn were extracted from these signals and classified using the linear discriminant analysis (LDA) [18, 19]. The leave-one-out (LOO) cross-validation was applied to test the accuracy of LDA-based classification of the two types of features. Figure 6.7 shows the areas under the receiver operating characteristic curves (ROC) [20], denoted as AUC (area under curve), obtained from the MSE and TSME features extracted from the PPG signals, where the AUC of the MSE = 0.74 and the AUC of the TSME = 0.84 (the higher the value of the AUC, the better the performance). Table 6.3 shows the sensitivity (the rate of elderly features that are correctly identified), specificity (the rate of caregiver features that are correctly identified), and accuracy rates obtained from the LDA using the MSE and TSME features. The TSME feature yielded better results than the MSE feature in terms of accuracy and receiver operating characteristics.

The first 5000 samples of the EMG right bicep signals of the four subjects in The Physical Action Data Set [17] were also used for differentiating the normal from aggressive actions, using LDA analysis with MSE and TSME features. The experiments were carried out on each subject performing ten normal and ten aggressive physical actions that represented human activities. The ten normal actions are (1) bowing, (2) clapping, (3) handshaking, (4) hugging, (5) jumping, (6) running, (7) seating, (8) standing, (9) walking, and (10) waving. The ten aggressive actions include (1) elbowing, (2) front kicking, (3) hammering, (4) heading, (5) kneeing, (6) pulling, (7) punching, (8) pushing, (9) side kicking, and (10) slapping.

Figure 6.8 shows the ROC areas under curves obtained from the MSE and TSME features, where the AUC of the MSE = 0.79 and the AUC of the TSME = 0.84, showing the better performance of the TSME. Table 6.4 shows the sensitivity (the rate of aggressive action features that are correctly identified), specificity (the rate of normal action features that are correctly identified), and accuracy rates obtained from the LDA using the MSE and TSME features. Once again, the TSME feature yielded better results than the MSE feature in general. The specificity (77.50%) obtained from the MSE is 5% higher than the specificity (72.50%) obtained from the TSME, while the sensitivity (67.50%) obtained from the MSE is 15% lower than the sensitivity (82.50%) obtained from the TSME, and the accuracy (50%) of the MSE is 11.25% lower than the accuracy (61.25%) of the TSME.

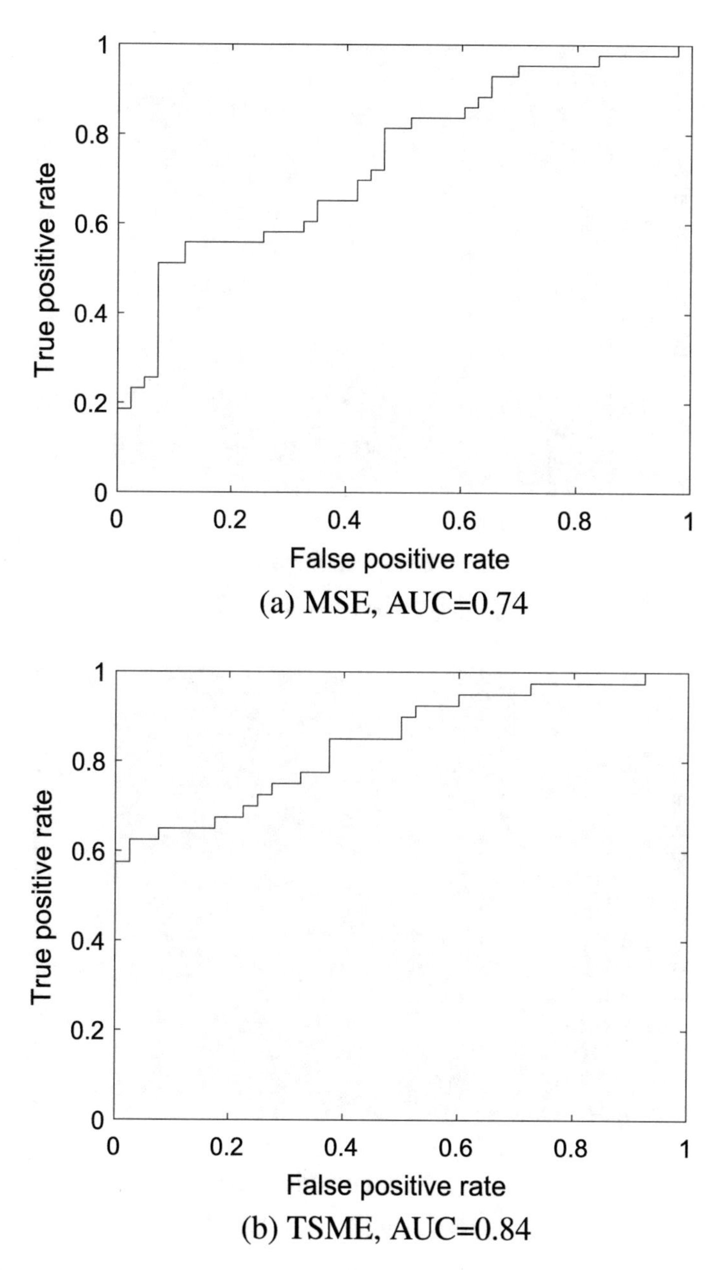


**Fig. 6.7** ROC curves obtained from MSE and TSME features of synchronized PPG signals of elderly participants and caregiver using linear discriminant analysis

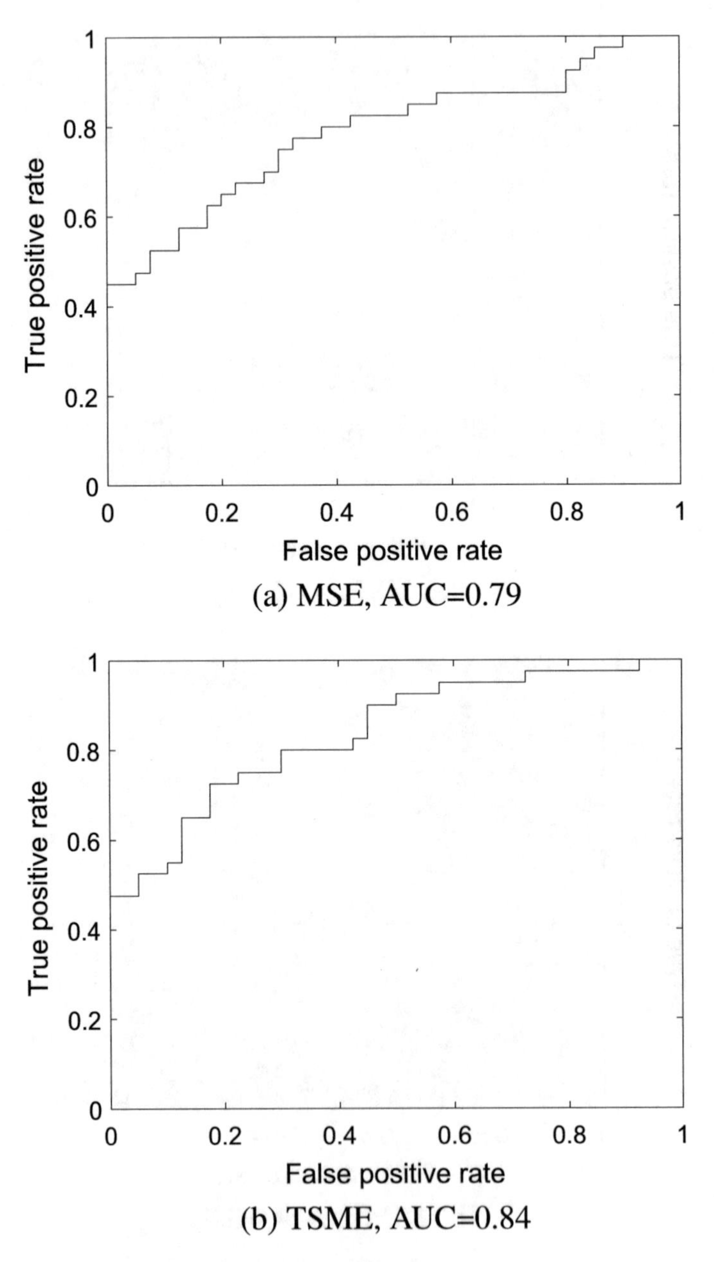


**Fig. 6.8** ROC curves obtained from MSE and TSME features of EMG signals of normal and aggressive actions using linear discriminant analysis

**Table 6.3** Sensitivity (SEN), specificity (SPE), and leave-one-out (LOO) cross-validation obtained from linear discriminant analysis of synchronized PPG signals of elderly participants and caregiver, using MSE and TSME

| Feature | SEN (%) | SPE (%) | LOO (%) |
|---|---|---|---|
| MSE | 51.16 | 93.02 | 44.19 |
| TSME | 62.50 | 97.50 | 62.50 |

**Table 6.4** Sensitivity (SEN), specificity (SPE), and leave-one-out (LOO) cross-validation obtained from linear discriminant analysis of EMG signals of normal and aggressive actions, using MSE and TSME

| Feature | SEN (%) | SPE (%) | LOO (%) |
|---|---|---|---|
| MSE | 67.50 | 77.50 | 50.00 |
| TSME | 82.50 | 72.50 | 61.25 |

### *6.4.3 Remarks*

SampEn has been reported to produce results better than the use of ApEn in several studies of time-series analysis [21–23]. The use of SampEn is more theoretically sound for computing TSME as it is adopted for computing MSE. The MSE has been found useful in many applications, most recently such as [24–27].

The construction of the new time series for computing the entropy profiles based on the HFD was known to be able to provide stable time scale and indices corresponding to the characteristics of irregular time series, including short time series. This stability can be achieved by taking into account self-similarity across the characteristic time scale [6]. While the MSE uses the averaging of time series on several interval scales, the TSME applies time shifting in time series that is based on the calculation of the "mean length" of the curve of a time series implemented in the HFD. The examples using noise and chaotic time series suggest the validity of the TSME, and classification results using the physiological data illustrate the better performance of the TSME over the MSE.

The computational time required for computing the TSME is higher than for computing the MSE, particularly when $k_{max}$ is large. Being similar to the specification of the $\tau$ parameter used in the MSE, an optimal selection for the $k_{max}$ used in the TSME is still an ad hoc choice reported in many studies [11], which needs further investigation. Future development of the TSME for multivariate multiscale entropy is worth pursuing as it has been formulated based on the concept of the MSE [28, 29]. An extension of the differential Shannon entropy rate of time series using kernel density estimators for selecting the order and bandwidth parameters [30] may have implication for improving the performance of the TSME .

An attempt to reduce the computational cost for computing TSME by means of ApEn and SampEn was proposed by identifying dissimilar vectors that can be excluded in the entropy computation to increase computing speed [31]. This can be particularly useful for processing large datasets. An extension of the TMSE by combining the TSME with the fuzzy entropy (FuzzyEn) [32], which is called time-shift multiscale fuzzy entropy (TSMFE) [33], has recently been proposed to overcome some limitations of the MSE. The use of the theory of geostatistics for computing entropy of pattern complexity and similarity in time series was also proposed in literature [34].

## References


1. Pham TD (2017) Time-shift multiscale entropy analysis of physiological signals. Entropy 19:257
2. Pincus SM (1991) Approximate entropy as a measure of system complexity. Proc Natl Acad Sci USA 88:2297–2301
3. Pincus SM, Gladstone IM, Ehrenkranz RA (1991) A regularity statistic for medical data analysis. J Clin Monit 7:335–345
4. Richman JS, Moorman JR (2000) Physiological time-series analysis using approximate entropy and sample entropy. Am J Physiol Heart Circ Physiol 278:H2039–H2049
5. Costa M, Goldberger AL, Peng CK (2002) Multiscale entropy analysis of complex physiologic time series. Phys Rev Lett 89:068102
6. Higuchi T (1988) Approach to an irregular time series on the basis of the fractal theory. Phys D 31:277–283
7. Higuchi T (1990) Relationship between the fractal dimension and the power law index for a time series: a numerical investigation. Phys D 46:254–264
8. Spasic S et al (2008) Spectral and fractal analysis of cerebellar activity after single and repeated brain injury. Bull Math Biol 70:1235–1249
9. Spasic S et al (2011) Different anaesthesia in rat induces distinct inter-structure brain dynamic detected by Higuchi fractal dimension. Fractals 19:113–123
10. Klonowski W (2009) Everything you wanted to ask about EEG but were afraid to get the right answer. Nonlinear Biomed Phys 3:2
11. Kesic S, Spasic SZ (2016) Application of Higuchi's fractal dimension from basic to clinical neurophysiology: a review. Comput Methods Progr Biomed 133:55–70
12. Steeb WH (2015) The nonlinear workbook. World Scientific, Singapore
13. Carter B (2013) Op amps for everyone, 4th edn. Elsevier, MA (USA)
14. Ward LM, Greenwood PE (2007) $1/f$ noise. Scholarpedia 2:1537
15. Lorenz EN (1963) Deterministic nonperiodic flow. J Atmos Sci 20:130–141
16. Pham TD et al (2015) Computerized assessment of communication for cognitive stimulation for people with cognitive decline using spectral-distortion measures and phylogenetic inference. PLoS ONE 10:e0118739
17. Lichman M (2013) UCI machine learning repository. http://archive.ics.uci.edu/ml. Accessed 08 Sept 2016
18. Fisher RA (1936) The use of multiple measurements in taxonomic problems. Ann Eugen 7:179–188
19. McLachlan GJ (2004) Discriminant analysis and statistical pattern recognition. Wiley-Interscience, New York
20. Metz CE (1978) Basic principles of ROC analysis. Semin Nucl Med 8:283–298
21. Al-Angari HM, Sahakian AV (2007) Use of sample entropy approach to study heart rate variability in obstructive sleep apnea syndrome. IEEE Trans Biomed Eng 54:1900–1904
22. Alcaraz R, Rieta JJ (2010) A review on sample entropy applications for the non-invasive analysis of atrial fibrillation electrocardiograms. Biomed Signal Process Control 5:1–14

23. Rostaghi M, Azami H (2016) Dispersion entropy: a measure for time-series analysis. IEEE Signal Process Lett 23:610–614
24. Humeau-Heurtier A (2015) The multiscale entropy algorithm and its variants: a review. Entropy 17:3110–3123
25. Grandy TH et al (2016) On the estimation of brain signal entropy from sparse neuroimaging data. Sci Rep 6:23073
26. Busa MA, van Emmerik REA (2016) Multiscale entropy: a tool for understanding the complexity of postural control. J Sport Health Sci 57:44–51
27. Stosic D et al (2016) Correlations of multiscale entropy in the FX market. Phys A 457:52–61
28. Humeau-Heurtier A (2016) Multivariate generalized multiscale entropy analysis. Entropy 18:411
29. Ahmed MU et al (2017) A multivariate multiscale fuzzy entropy algorithm with application to uterine EMG complexity analysis. Entropy 19:2
30. Darmon D (2016) Specific differential entropy rate estimation for continuous-valued time series. Entropy 18:190
31. Lu Y et al (2017) Accelerating the computation of entropy measures by exploiting vectors with dissimilarity. Entropy 19:598
32. Chen W et al (2007) Characterization of surface EMG signal based on fuzzy entropy. IEEE Trans Neural Syst Rehabil Eng 15:266–272
33. Zhu X et al (2018) Time-shift multiscale fuzzy entropy and Laplacian support vector machine based rolling bearing fault diagnosis. Entropy 20:602
34. Pham TD (2010) GeoEntropy: a measure of complexity and similarity. Pattern Recognit 43:887–896

# Chapter 7
# Applications in Biomedicine

## 7.1 Protein Expression by Immunohistochemistry in Rectal Cancer Patients

Colorectal cancer (CRC) is the third most common cancer in the world [1]. There are around 6500 new diagnosed cases of CRC yearly among the population of Sweden. The causes for CRC are thought to be associated with gene mutations, gene variants, and changed expression of proteins. The combination of surgery and radio-chemotherapy is the most beneficial regimens in current treatment of advanced rectal cancer. Preoperative radiotherapy (RT) is often given to rectal cancer patients as a complement to surgery to improve treatment outcome. However, tumor recurrence plays a major cause of death for progressive rectal patients after surgery. As a result, a significant proportion of patients did not benefit from preoperative RT [2].

It remains up to date that the clinical testing of specific mutations in KRAS, BRAF, RAS, and RAF genes along with mismatch repair gene deficiency assists either as prognostic or predictive biomarkers in CRC [3, 4]. Other methods for identifying biomarkers in the treatment of CRC include molecular subtype classification [5], identifying molecular signatures at protein and RNA levels by microarray analysis [6], protein identification in cell proliferation and new blood vessels [7], changes in the amounts of certain proteins [8], and proteomic strategies [9]. A recent review of methods for discovering prognostic and predictive biomarkers in CRC for personalized therapy can be found in [10].

The role of predictive biomarkers is known to be essential for the field of radiation oncology [11]. While most efforts aim to improve cancer treatment with respect to physical conditions and technology, such as precision in treatment plans and dose administration [12], the inclusion of patient-specific biological characteristics into cancer treatment decision would be very useful for personalized treatment. However, such important biological information of individual patients is not well explored. To achieve this purpose, predictive biomarkers are needed to guide radiation oncologists to determine optimal dose prescription, select patient-specific schemes, and treatments for individual cancer patients [11].



T. D. Pham, *Fuzzy Recurrence Plots and Networks with Applications in Biomedicine*, https://doi.org/10.1007/978-3-030-37530-0_7

The TP73 gene is located to the 1p36 locus and encodes for several isoforms with distinct functions. The expression of p73 protein in cancer is important for predicting prognosis and therapeutic response, which remains controversial. An important reason is that the TP73 gene expresses two different categories of proteins, full-length isoforms (TAp73) and N-terminal truncated variants (DNp73), which mainly arise due to the usage of two alternate promoters; the upstream P1 promoter generates TAp73 isoforms, whereas the downstream P2 promoter yields the DNp73 isoforms. Furthermore, DNp73 isoforms can be transcribed from P1 promoter, but due to alternative splicing and different initiation of translation, the protein is the same as the one transcribed from P2 promoter.

Oncogenesis includes ten steps or hallmarks, i.e., sustaining proliferative signaling, evading growth suppressors, activating invasion and metastasis, enabling replicative immortality, inducing angiogenesis, resisting cell death, tumor-promoting inflammation, genome instability, and reprogramming of energy metabolism (Fig. 7.1) [13].

The increased levels of TP73 mRNA transcription are found in various tumors compared with the surrounding normal tissues. The imbalance between TAp73 and DNp73 isoforms may be useful to predict response to chemotherapy and prognosis [14, 15]. High DNp73 expression has strong correlation with unfavorable prognosis in several types of cancer patients, and DNp73-positive tumors show a reduced response to chemotherapy and irradiation [16–18]. The upregulation of DNp73 was frequently detected in radioresistant cervical cancers [19]. Our previous findings indicated that DNp73 is increased in colon cancer cell line that is resistant to $\gamma$-irradiation [20]. Thus, these findings suggested that DNp73 expression may play an important role in the regulation of radiosensitivity. However, the prognostic and predictive role of DNp73 in rectal cancer patients with radiation still remains unclear.

This study was to elucidate the role of DNp73 as a predictive biomarker by investigating if DNp73 was related to the survival time of rectal cancer patients who were administered with RT before surgery [21]. To overcome the subjective and time-consuming task of pathologist-based analysis of immunohistochemistry (IHC) images stained for DNp73 expression, we carried out a study by applying the concept of MC-FWRNs described in Chap. 5. The motivation for using this new image-based network analysis is that the recurrence of image attributes inherently existing in the complex nature of IHC images of rectal cancer tissue arrays.

In fact, network analysis in graph theory has been increasingly recognized as a useful tool for studying cancer. Such studies include the prediction of outcomes of ovarian cancer treatment [22], analysis of breast cancer progression and reversal [23], drug response prediction in cancer cell lines [24], identification of novel cancer gene candidates [25], tumor biology for precision cancer medicine [26], and prediction of cancer recurrence [27].

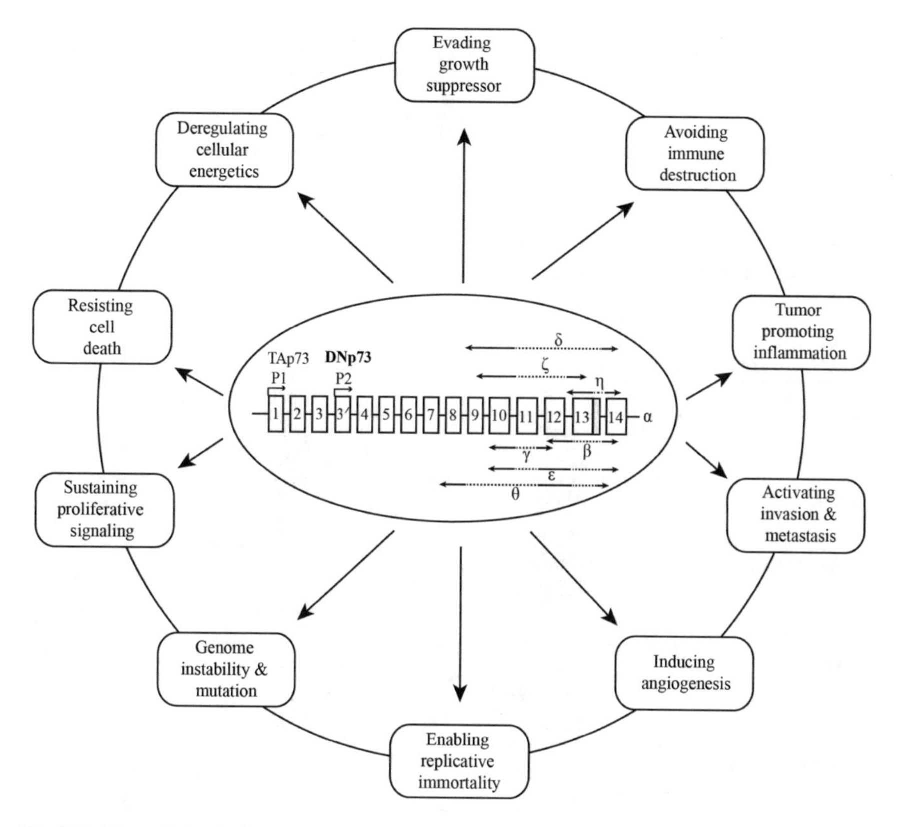


**Fig. 7.1** The splicing isoforms and function of p73 in cancer. TAp73 isoforms are transcribed from the external P1 promoter whereas DNp73 isoforms are transcribed from the internal P2 promoter. Alternative splicing after exon 9 gives rise to P73 C-terminal variants, $\beta$, $\gamma$, $\delta$, $\epsilon$, $\zeta$, $\eta$, and $\theta$ (dotted line means loss of corresponding exon). The TP73 plays multiple functions in cancer (modified from [13])

## *7.1.1 Rectal Cancer Patients*

This study included the patients with rectal adenocarcinoma from the Southeast Swedish Health Care region who participated in a clinical trial of preoperative RT for rectal cancer [28]. Samples of biopsy and primary tumor from the same patients were selected for the analysis. In this Swedish Rectal Cancer Trial study, we collected samples from both pre-radiotherapy and non-radiotherapy rectal cancer patients. The biopsy samples were taken from the rectal cancer before the RT and went through the routine pathological process, and eventually embedded in paraffin blocks. The primary tumor samples were taken from the primary rectal cancer after the RT.

There were 25 patients with RT whose demographic information is given in Table 7.1. This study was carried out in accordance with the recommendations of Good Clinical Practice, the Research Ethics Committee in Linkoping, Sweden with

**Table 7.1** Demographic information of the rectal cancer patients who had a median age of 68 years (range: 39–78 years) were followed for a median period of 81 months (range: 0–129 months) and had the median time to disease free of 101 months after surgery (range: 15–288 months)

| | Number of patients |
|---|---|
| Male | 16 (64%) |
| Female | 9 (36%) |
| Shorter survival time (15–75 months) | 11 (44%) |
| Longer survival time (101–288 months) | 14 (56%) |

**Table 7.2** Clinico-pathological characteristics of the rectal cancer patients

| Parameters | | Number of cases |
|---|---|---|
| Age | <60 | 6 |
| | >60 | 19 |
| Gender | Male | 9 |
| | Female | 16 |
| Growth pattern | Expansion | 11 |
| | Infiltration | 13 |
| | Null | 1 |
| Grade | Well | 2 |
| | Moderate | 14 |
| | Poor | 9 |
| Pathological stages | I | 8 |
| | II | 6 |
| | III | 8 |
| | IV | 3 |

written informed consent from all subjects. All subjects gave written informed consent in accordance with the Declaration of Helsinki. The protocol was approved by the Research Ethics Committee in Linkoping, Sweden. The clinico-pathologic characteristics of the patients are listed in Table 7.2.

### 7.1.2 *Immunochemistry and Image Extraction*

The five-micrometer paraffin-embedded tissue microarray (TMA) sections were deparaffinized in xylene and rehydrated with a series of gradient ethanol to water. The sections were heated to boiling point in citrate buffer (pH 6.0) for 30 min to unmasked antigen, followed by a washing in phosphate-buffered saline (PBS). Endogenous peroxidase activity was blocked with 3% $H_2O_2$ in methanol followed by washing three times in PBS. The sections were incubated with protein block (Dako, Carpinteria,

CA) for 10 minutes and then incubated with anti-DNp73 antibody (clone 38C674.2, Novus Biologicals, 1:200), which specifically recognized all DNp73 isoforms, but not TAp73.

After that, the sections were washed in PBS and then incubated with goat anti-mouse secondary antibody (Dako) at room temperature for 25 min. Next, the sections were subjected to 3,3'-diaminobenzidine tetrahydrochloride for 8 min and then counterstained with hematoxylin. Negative and positive controls were added in each staining run. All slides were scored by two independent investigators. Whole-slide images of entire sections were captured with an Aperio CS2 slide scanner system (Leica Biosystems, Wetzlar, Germany) using a 40x magnification.

All sections were reviewed to remove images containing tissue-processing artifacts, including bubbles, section folds, and poor staining. A total of 46 whole-slide images from the 25 unique patients were extracted from the TMA slides using ObjectiveViewer (https://www.objectivepathology.com/objectiveview) with the original resolution.

### *7.1.3 Results*

Table 7.3 shows the screening results of the 25 rectal cancer patients. Patient numbers 1–11 are those who had shorter survival time, and patient numbers 12–25 are those who had longer survival time. The evaluation of the IHC-stained color intensity of the whole slide of a tissue core with brown antibody stain and blue counterstain was assessed as being positive and negative, respectively. The positive stain is subjectively classified as weak = 1 (light brown), moderate = 2 (moderate brown), and strong = 3 (dark brown), whereas the negative stain = 0 (blue). Figure 7.2 shows representative IHC staining for DNp73 expression on the biopsy and primary tumor tissue images obtained from a rectal cancer patient survived 40 months after radiotherapy, and biopsy and primary tumor tissue images obtained from a rectal cancer patient who survived 255 months after radiotherapy at the censoring date.

To capture the local information of the DNp73 expression over the whole IHC-stained slides, images of biopsy and primary tumor of each of the 25 rectal cancer patients were divided into subimages of $150 \times 150$ pixels. The subimages that contain either the background or a large portion of the background were excluded in the analysis. To construct the FWRNs of the IHC-stained subimages, we selected the FWRN parameters $m = 3$ to establish a reasonable local window size of $7 \times 7$, $c = 20$ that was approximately based on the partition entropy, and the FCM parameters: weighting exponent = 2, maximum number of iteration = 100, and convergence rate tolerance = 0.00001, which are widely adopted for the FCM analysis. The clustering coefficient and characteristic path length were calculated for each subimage of each patient, and the total average values of the clustering coefficients and characteristic path lengths of all subimages represent the reported values.

Figures 7.3 and 7.4 show the clustering coefficients and characteristic path lengths of the FWRNs of the biopsy and primary tumor images obtained from the 25 rectal

**Table 7.3** Screening results of rectal cancer patients

| Patient # | Disease-free time | Recurrence status | Survival time | IHC score | |
|---|---|---|---|---|---|
| | | | | Primary tumor | Biopsy |
| 1 | 0 | Yes | 15 | 1 | 1 |
| 2 | 6 | Yes | 19 | 1 | 3 |
| 3 | 20 | Yes | 25 | 2 | 1 |
| 4 | 13 | Yes | 40 | 3 | 3 |
| 5 | 37 | Yes | 60 | 3 | 3 |
| 6 | 44 | Yes | 62 | 2 | 3 |
| 7 | 0 | Yes | 15 | 3 | 2 |
| 8 | 63 | Yes | 75 | 3 | 3 |
| 9 | 26 | Yes | 27 | 3 | 3 |
| 10 | 12 | No | 26 | 3 | 3 |
| 11 | 34 | Yes | 43 | 2 | 2 |
| 12 | 100 | Yes | 101 | 1 | 1 |
| 13 | 0 | Yes | 180 | 1 | 3 |
| 14 | 114 | Yes | 114 | 2 | 2 |
| 15 | 122 | Yes | 255 | 3 | 2 |
| 16 | 167 | Yes | 167 | 2 | 2 |
| 17 | 81 | No | 101 | 3 | 2 |
| 18 | 129 | Yes | 129 | 2 | 3 |
| 19 | 129 | Yes | 288 | 2 | 3 |
| 20 | 126 | Yes | 126 | 3 | 2 |
| 21 | 186 | Yes | 238 | 3 | 3 |
| 22 | 122 | Yes | 122 | 2 | 2 |
| 23 | 168 | Yes | 288 | 2 | 3 |
| 24 | 151 | Yes | 151 | 3 | 3 |
| 25 | 168 | Yes | 168 | 2 | 2 |

*Note* Time is in months. For IHC score, 0 = negative, 1 = weak, 2 = moderate, and 3 = strong

cancer patients, respectively. The scatter plot of the survival time against the ratios of the clustering coefficients of the primary tumors to those of the biopsies, and the ratios of the characteristic path lengths of the primary tumors to those of the biopsies are shown in Fig. 7.5.

Based on the visualization of the scatter plot, we discovered the predictive value of DNp73 in the rectal cancer patients in terms of the clustering-coefficient and characteristic-path-length ratios, which are shown in Fig. 7.6. The probability (p) for the predicted survival time based on the clustering-coefficient ratio was computed as the number of patients who lived between 101 and 288 months divided by the total number of patients whose clustering-coefficient ratios are within the ratio range ($p = 11/15 = 0.7333$). The probability (p) for the predicted survival time based

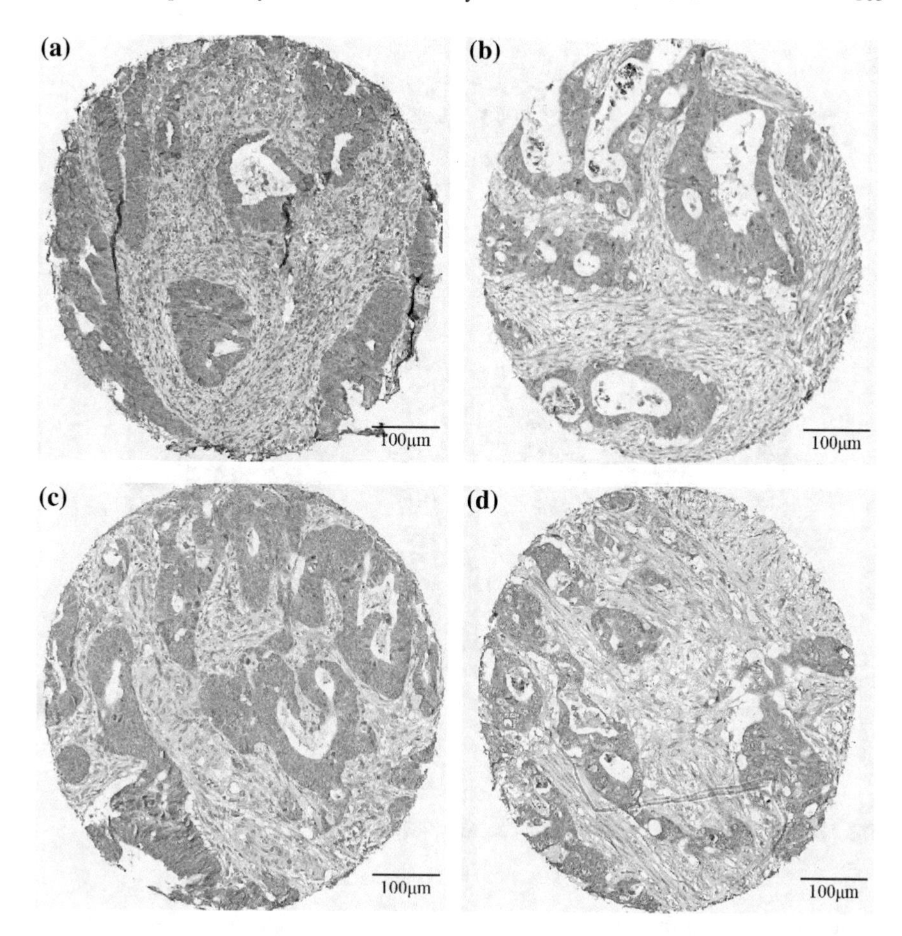


**Fig. 7.2** Representative IHC-stained images of DNp73 expression: **a** a biopsy image, **b** a primary tumor image obtained from a rectal cancer patient survived 40 months after radiotherapy, **c** a biopsy image, and **d** a primary tumor image obtained from a rectal cancer patient survived 255 months after radiotherapy

on the characteristic-path-length ratio was computed as the number of patients who lived between 126 and 288 months divided by the total number of patients whose characteristic-path-length ratios are within the ratio range ($p = 7/9 = 0.7778$).

The findings presented herein show the useful application of complex network analysis of images for studying the predictive factor of DNp73 biomarker expression in rectal cancer patients. The use of DNp73 biomarker can give insight into preoperative RT that has been considered as an important companion in the treatment of rectal cancer. A larger sample size when being available in future clinical trial will further confirm the current findings. Moreover, the proposed approach is not only found useful to rectal cancer but also can be adopted for the analysis of other biomarkers as well as other types of cancer, where human-based pathology practice is of

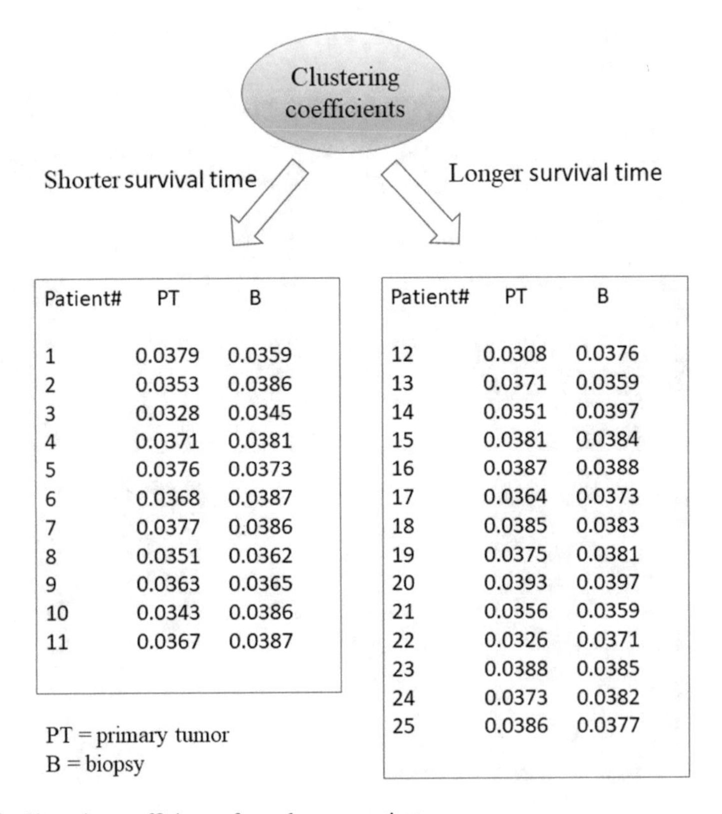


| Patient# | PT | B |
|---|---|---|
| 1 | 0.0379 | 0.0359 |
| 2 | 0.0353 | 0.0386 |
| 3 | 0.0328 | 0.0345 |
| 4 | 0.0371 | 0.0381 |
| 5 | 0.0376 | 0.0373 |
| 6 | 0.0368 | 0.0387 |
| 7 | 0.0377 | 0.0386 |
| 8 | 0.0351 | 0.0362 |
| 9 | 0.0363 | 0.0365 |
| 10 | 0.0343 | 0.0386 |
| 11 | 0.0367 | 0.0387 |

| Patient# | PT | B |
|---|---|---|
| 12 | 0.0308 | 0.0376 |
| 13 | 0.0371 | 0.0359 |
| 14 | 0.0351 | 0.0397 |
| 15 | 0.0381 | 0.0384 |
| 16 | 0.0387 | 0.0388 |
| 17 | 0.0364 | 0.0373 |
| 18 | 0.0385 | 0.0383 |
| 19 | 0.0375 | 0.0381 |
| 20 | 0.0393 | 0.0397 |
| 21 | 0.0356 | 0.0359 |
| 22 | 0.0326 | 0.0371 |
| 23 | 0.0388 | 0.0385 |
| 24 | 0.0373 | 0.0382 |
| 25 | 0.0386 | 0.0377 |

**Fig. 7.3** Clustering coefficients of rectal cancer patients

limited capacity. In fact, there are many reports on the computerized image analysis of H&E (Haematoxylin and Eosin) staining, much less effort has been made to apply computational methods for the automated analysis of IHC staining. The MC-FWRN presented in this paper can be generally applied for studying the expression of other potential biomarkers. Furthermore, the investigation of other properties of complex networks of DNp73 expression on histology images of rectal cancer is worth being carried out in future research to discover more image-driven associations between RT and patient outcome, which can be promising as important information on the likelihood of response and potential tools in personalized oncology.

Although there are many studies reported about the association between DNp73 protein biomarker expression and malignant potential, the function of DNp73 still remains unclear. This work contributes to the elucidation of the predictive value of DNp73 expression in rectal cancer patients who were given preoperative RT. We developed an original method for constructing weighted recurrence networks of multichannel images. These networks allow the extraction of useful network properties from complex IHC images. The clustering coefficients and characteristic path lengths of the MC-FWRNs are not only able to show the predictive factor of DNp73 expres-

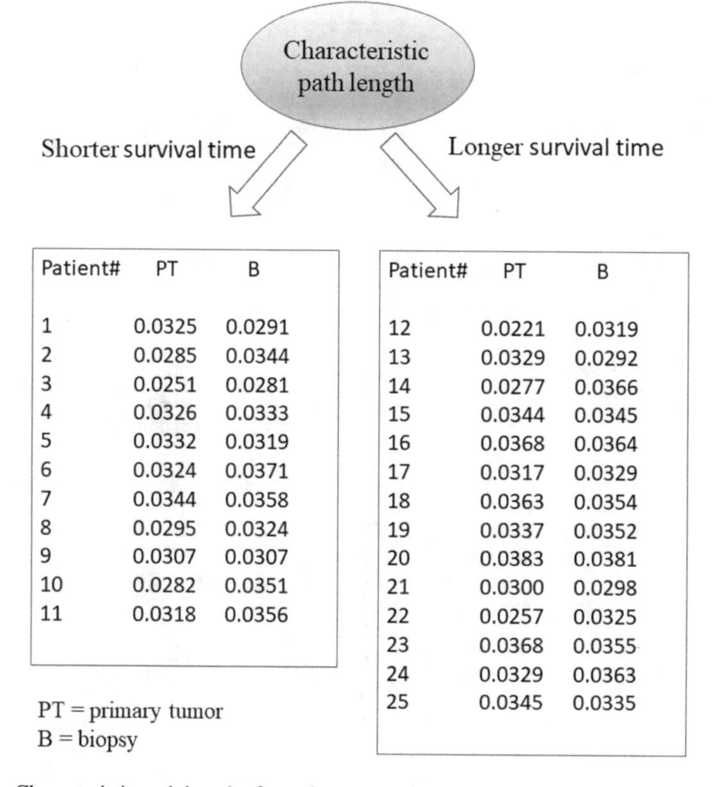


| Patient# | PT | B |
|---|---|---|
| 1 | 0.0325 | 0.0291 |
| 2 | 0.0285 | 0.0344 |
| 3 | 0.0251 | 0.0281 |
| 4 | 0.0326 | 0.0333 |
| 5 | 0.0332 | 0.0319 |
| 6 | 0.0324 | 0.0371 |
| 7 | 0.0344 | 0.0358 |
| 8 | 0.0295 | 0.0324 |
| 9 | 0.0307 | 0.0307 |
| 10 | 0.0282 | 0.0351 |
| 11 | 0.0318 | 0.0356 |

| Patient# | PT | B |
|---|---|---|
| 12 | 0.0221 | 0.0319 |
| 13 | 0.0329 | 0.0292 |
| 14 | 0.0277 | 0.0366 |
| 15 | 0.0344 | 0.0345 |
| 16 | 0.0368 | 0.0364 |
| 17 | 0.0317 | 0.0329 |
| 18 | 0.0363 | 0.0354 |
| 19 | 0.0337 | 0.0352 |
| 20 | 0.0383 | 0.0381 |
| 21 | 0.0300 | 0.0298 |
| 22 | 0.0257 | 0.0325 |
| 23 | 0.0368 | 0.0355 |
| 24 | 0.0329 | 0.0363 |
| 25 | 0.0345 | 0.0335 |

**Fig. 7.4** Characteristic path length of rectal cancer patients

sion in the patients, but also reveal the identification of noneffective application of RT to those who had poor overall survival outcome.

Both intensity and percentage of the IHC staining must be considered when we score the slides. We have been working with such a classic scoring system for many years. We have realized that even two experienced pathologists score the slides, there is still difficulty to make clear decisions for about 10% of the cases. In this study, a new image-based network analysis was developed to analyze the immunostaining array slides and to extract patterns of the IHC staining, including both intensity and percentage in the whole arrays. We further analyzed the associations of the immunostaining patterns with our clinical data to provide more precise information for rectal cancer.

The mean values of both clustering coefficients and characteristic path lengths of the rectal cancer patients of shorter survival are lower than those of longer survival. There is no correlation between the ratios of the clustering coefficients of the tumor to those of the biopsy and the survival time (correlation coefficient $R = 0.0120$, p-value $= 9.7656e\text{-}04$) among the shorter surviving rectal cancer patients whose maximum survival time was about over 6 years (75 months). This can be observed from Fig. 7.5.

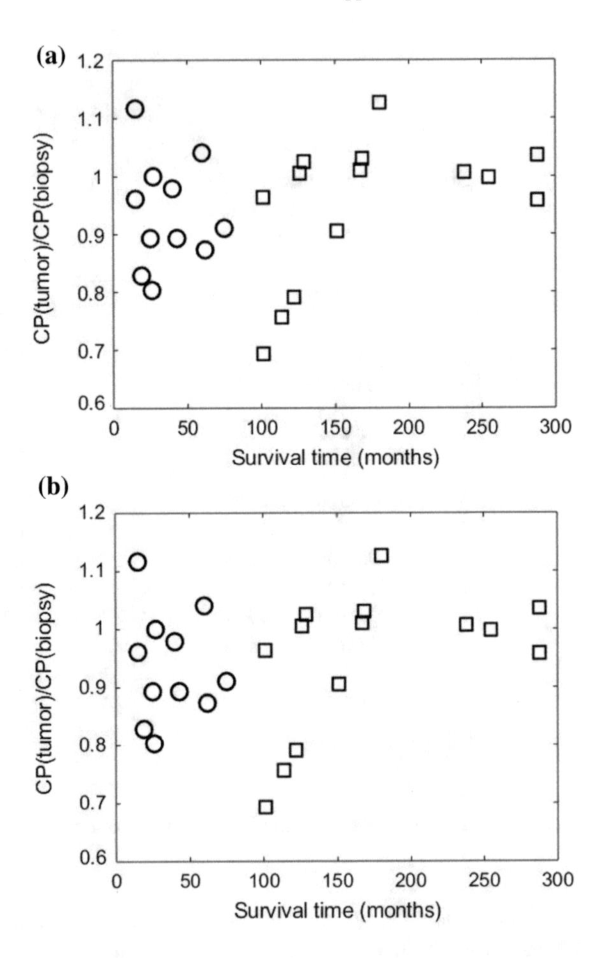


**Fig. 7.5** Scatter plots of survival time of 25 rectal cancer patient against **a** ratios of clustering coefficients of primary tumor (CC(tumor)) to those of biopsy (CC(biopsy)) and **b** ratios of characteristic path lengths of primary tumor (CP(tumor)) to those of biopsy (CP(biopsy)). Symbols "o" and "□" indicate patient groups with shorter and longer times of survival, respectively

There is also no correlation between the ratios of the characteristic path lengths of the tumor to those of the biopsy and the survival time (R = –0.0780, p-value = 9.7656e-04) among the shorter surviving patients. This can also be observed from Fig. 7.5. There is an indication of correlation between the ratios of the clustering coefficients of the tumor to those of the biopsy and the survival time (R = 0.4924, p-value = 1.2207e-04) among the longer surviving rectal patients whose maximum survival time was 24 years (288 months). This can be observed from Fig. 7.5. There is also evidence of correlation between the ratios of the characteristic path lengths of the tumor to those of the biopsy and the survival time (R = 0.4778, p-value = 1.2207e-04) among the longer surviving rectal cancer patients. This can also be observed from Fig. 7.5.

Figure 7.5 shows similar plots of the ratios of the two network-property parameters of the tumor to biopsy against the survival time, suggesting the consistency of the results. It is reported that rectal patients who survive at least 5 years (60

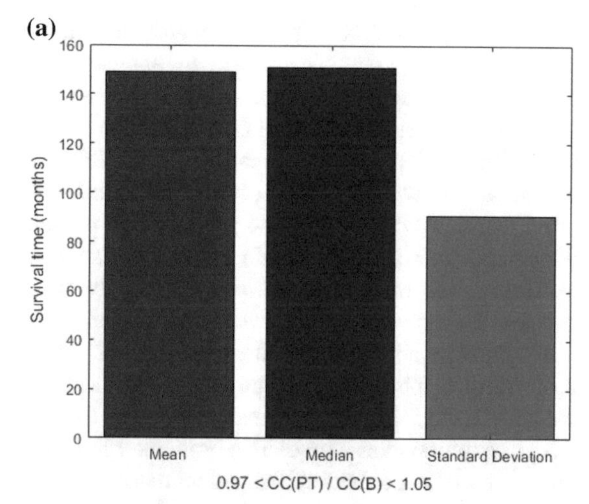


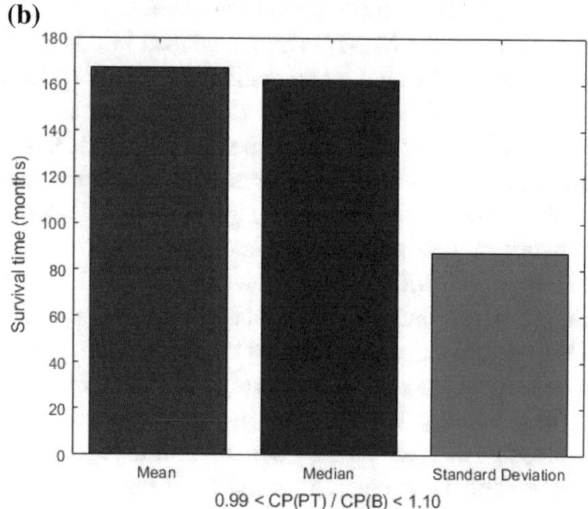


**Fig. 7.6** Predictive value of DNp73 in rectal cancer patients in terms of ratios of clustering coefficient (**a**) and characteristic path length (**b**) of image-based FWRNs. CC(PT) and CC(B) denote clustering coefficients of primary tumor and biopsy, respectively; and CP(PT) and CP(B) denote characteristic path lengths of primary tumor and biopsy, respectively

months) are likely to die from causes that are common in the general population [29]. This finding highlights the predictive value of DNp73 revealed by the image-based FWRN analysis among the cohort of rectal cancer patients whose survival time was between 8.4 years (101 months) and 24 years (288 months) correlated with the clustering-coefficient ratios, and 10.5 years (126 months) and 24 years (288 months) correlated with the characteristic-path-length ratios.

The lack of correlation of the ratios of the MC-FWRN parameters and the (shorter) survival time may suggest an implication of poor responses or noneffective treatment of the RT provided to the rectal cancer patients. Meanwhile, those patients who have positive correlation between the ratios of the FWRN parameters and the (longer) survival time were very likely to have a good or better response to the RT. General findings are that higher values of the ratios of the MC-FWRN parameters indicate longer survival time. The longest survival time is found with the values of the MC-FWRN parameter ratios being about 1. Based on the MC-FWRN parameters of the 25 rectal cancer patients and their survival months, we can predict the survival time between 101 months (8.42 years) and 288 months (24 years) with a probability of 73% for those patients whose clustering-coefficient ratio is within the range between 0.97 and 1.05. Similarly, the survival time between 126 months (10.5 years) and 288 months (24 years) with a probability of 78% for those patients whose characteristic-path-length ratio is within the range between 0.99 and 1.10.

The application of a novel image-based network analysis presented in this study was able to discover the predictive factor of DNp73 biomarker in rectal cancer patients having preoperative RT. Predictive biomarkers provide useful information on the probability of obtaining a response to treatment [30] and support the process of therapeutic decision for personalized cancer treatment [31]. Such a discovery of DNp73 expression as a predictive biomarker in rectal cancer patients is expected to provide early assessment of the patient outcome, clinical value in the diagnostics of the disease, and identification of targeted postoperative therapy.

Regarding the MC-FWRN introduced in this study, this new method appears to be the first of its kind mathematically formulated to capture the recurrence features of multichannel data inherently existing in complex histology images in a way that is both effective and easily implemented for practical use. Complex networks consist of certain attributes that can be computed to analyze the properties and characteristics of the networks. Mathematical properties of these networks are utilized to define network models and to elucidate how certain models are different to each other. The proposed MC-FWRN allows the calculation of the clustering coefficients and characteristic path lengths of DNp73 expression in the primary tumors and biopsies. These values can used to predict the survival time of a cohort of rectal cancer patients who were deemed to be positively influenced by preoperative RT.

In this study, we have shown significant results concerning the DNp73 protein expression in predicting the outcome for the rectal cancer patients with the proposed mathematical approach. A limitation of this study is a relatively small number of the rectal cancer patients selected in the analysis. Therefore, future studies with more subgroups of rectal patients will be considered. Furthermore, results from rectal cancer patients with and without preoperative RT will be obtained and compared. Images of biopsies, primary cancers, and metastatic cancers should be further investigated. Eventually, we will analyze the associations of the reactions from tumor invasive margins and stroma with the patients' prognosis.

Another limitation in this study is that the TAp73 expression was not performed in the present 25 pairs of rectal cancer samples. It is known that TAp73 acts as a tumor suppressor, while DNp73 exerts as an oncogene that is opposite to TAp73 [32, 33]. Therefore, it is necessary to expand the sample size and simultaneously evaluate TAp73 and DNp73 in the future, based on the methodology we have developed in the current study.

The highlights of the technical development and findings addressed in this paper are summarized as follows. First, the proposed MC-FWRN analysis of DNp73 expression by IHC in rectal cancer is the first of its kind. Second, a new mathematical analysis of IHC-stained biopsy and tumor images reveals the predictive power of DNp73 in rectal cancer patients who received RT. Third, a new method of multichannel fuzzy weighted recurrence networks is developed for extracting two useful complex network properties of IHC images that can be used as prognostic indicators of rectal cancer. Fourth, the proposed approach for quantifying the expression of IHC is not limited to the study of DNp73 but can also be generally applied to discovering image patterns of other tumor proteins. Fifth, the proposed approach can be utilized as a computerized tool for extracting features from whole-slide images in digital pathology.

## 7.2 The Recurrence Dynamics of Personalized Depression

Depression, including major depressive disorder or clinical depression, is a common but serious mood disorder. For many people with depression, symptoms usually are severe enough to cause noticeable problems in handling daily activities, such as feeling, thinking, sleeping, eating, working, school, social activities, or personal relationships. Symptoms caused by major depression can vary from person to person. For some people with major depression, they may feel generally miserable or unhappy without knowing the cause [34, 35]. Therefore, depression has significant global economic and social impacts [36]. Untreated depression increases the chance of risky behaviors such as drug or alcohol addiction, harming relationships, causing problems at work, and making it difficult to overcome serious illnesses [37]. Particularly, it is well recognized that late-life depression can result in significant morbidity and mortality but often goes undetected or untreated [38].

Recent literature has pointed out that mental disorders such as major depression involves continuous gene–environment interactions that exhibit behaviors in highly personalized manners, and the theory of complex system dynamics is promising as a useful tool for providing answers to the assessment of personalized risk for transitions on the continuum of depression [39]. Complex dynamical systems have critical transitions or thresholds known as tipping points at which the systems change rapidly from one state to another, suggesting the reality of early warning signals if the occurrence of a critical event is imminent [40]. Research findings support the hypothesis that mood may change into different states marked with tipping points and depression with its mood dynamics behaves like a complex system that sudden rises and gains are frequently observed [39, 41].

As a related methodology, complex networks have also been recognized as promising techniques for investigating depression and anxiety symptom relationships [42], prognosis of treatment response in depression [43], modeling mental disorders [44], and discovering a prognostic marker of treatment response in adolescent depression [45].

Being motivated by the potential of methods for analysis of the dynamics of complex systems for assessing depression, this paper presents an approach in chaos theory for studying recurrences of dynamical systems and complex networks that can offer gaining insight into the transition of depression under the gradual reduction of medication, recorded from complex time series over a large scale. The rest of this paper includes the description of a time-series database for major depression, mathematical methods of fuzzy recurrence analysis, fuzzy recurrence networks, and tensor decomposition for exploring latent patterns of the depression data, results, discussion, and concluding remarks of the research findings.

### 7.2.1 Time-Series Data

The time series obtained from a single participant used in this study are the data from "Critical Slowing Down as a Personalized Early Warning Signal for Depression" [39]. The participant is a 57-year-old male who was diagnosed with major depressive disorder and had been using antidepressants for 8.5 years [46] at the time of the experiment.

The background, methods, description, reuse potential, and public availability of the data were reported in [47]. The dataset contains around 1500 measurements and almost 50 items, which were completed at different time scales on momentary, daily, and weekly levels. As being pointed out by Kossakowski et al. [47], the time-series data are of a large scale that can be used for several purposes, and particularly very suitable for validating new methods for predicting the onset of a critical transition in personalized major depression. Furthermore, the dataset can be used to construct a network of a dynamical system and study how the network changes over time.

The participant completed 1478 measurements over the course of 239 consecutive days in 2012 and 2013. The entire study looked at momentary affective states in daily life before, during, and after the double-blind phase. The items, which were asked ten times a day, cover topics like mood, physical condition, and social contacts. Also, depressive symptoms were measured on a weekly basis using the Symptom Checklist Revised (SCL-90-R).

The dataset includes five main components of the mental states that are constructed from the time series of 12 affective items: (1) restless, (2) agitated, (3) irritated, (4) anxious, (5) lonely, (6) guilty, (7) enthusiastic, (8) cheerful, (9) content, (10) strong, 11) worrying, and 12) suspicious. The five mental states are (1) unrest, (2) negative, (3) positive, (4) worrying, and (5) suspicious. The unrest state includes restless, agitated, irritated, and anxious. The negative affect consists of lonely and guilty moods. The positive affect includes enthusiastic, cheerful, content, and strong moods. Worrying and suspicious moods are the other two mental states. These items were measured on a 7-point Likert scale, ranging from 1 (not) to 7 (very). The items concerned with feeling down, lonely, anxious, and guilty were measured on a 7-point Likert scale, ranging from –3 (not) to 3 (very). The data collection was carried out with five phases. Phase 1 is a baseline measurement period that lasted 4 weeks. Phase 2 is a double-blind period in which the antidepressant dosage was not yet reduced, which lasted between 0 and 6 weeks. Phase 3 is a double-blind period in which the antidepressant dosage was gradually reduced from 150 mg (venlafaxine) to 0 mg, which lasted 8 weeks. Phase 4 is a post-assessment period in which the antidepressant dosage was not changed, which lasted 8 weeks. Phase 5 is a follow-up period that lasted 12 weeks.

The participant wished to obtain an insight into his personal behavior of depression during a period in which the gradual reduction of antidepressants was performed. The participant aimed to discover if he would become more vulnerable to develop a new event of depression when the antidepressants were reduced, and whether this vulnerability could be detected in the data [47].

As stated in [47], "The participant (the 2nd author of this paper) initiated the study and expressed that he wanted the data to be published. Approval from the Maastricht University ethical committee was therefore unnecessary and not obtained. The participant gave his consent for collecting and (re)using the data."

### 7.2.2 Tensor Decomposition of Mental-State Dynamics

As a generalization of vector-based (one-way) or matrix-based (two-way) methods, tensor-based (multi-way) analysis is a powerful approach for modeling multidimensional data and extracting latent features for gaining insight into the underlying structure of the data [48]. The data in this study can be modeled as a three-mode tensor that is arranged in the following way: $Subject \times Mental\ states \times Recurrence\ dynamics$, where the number of subject $= 1$ (single participant), number of mental states $= 5$ (unrest, negative affect, positive affect, worry, and suspicious), and the recurrence dynamics consists of vectorized FJRPs or FRPs of the moods defining the corresponding mental states.

The PARAFAC (parallel factor analysis) [49, 50], which is a generalization of the principal component analysis (PCA) to arrays of higher orders, was applied in this study for computing the tensor-decomposition factors of the data. In tensor analysis, the terms *mode*, *way*, and *order* generally have the same meaning, and also the terms *component* and *factor* are used interchangeably [50]. In general, the PARAFAC model of an $N$-mode tensor of dimensions $L_j$, $j = 1, \ldots, N$, can be mathematically expressed as [51]

$$\underline{\mathbf{X}} = \sum_{f=1}^{F} \mathbf{a}_f^{(1)} \otimes \mathbf{a}_f^{(2)} \otimes \cdots \otimes \mathbf{a}_f^{(N)} + \underline{\mathbf{E}}, \tag{7.1}$$

where $\underline{\mathbf{X}} \in \mathcal{R}^{L_1 \times L_2 \times \cdots \times L_N}$ is the tensor, $\underline{\mathbf{E}}$ is the model error tensor, $\otimes$ is the outer product, $F$ is the number of factors, and $\mathbf{a}_f^{(j)}$, $j = 1, \ldots, N$, are the $f$th columns of the loading matrices $\mathbf{A}^{(j)} = [\mathbf{a}_1^{(j)}, \ldots, \mathbf{a}_F^{(j)}]$.

Alternatively, the tensor model can also be defined as

$$\underline{\mathbf{X}} = \underline{\hat{\mathbf{X}}} + \underline{\mathbf{E}} \approx \underline{\hat{\mathbf{X}}}, \tag{7.2}$$

where $\hat{\underline{\mathbf{X}}}$ is the estimate of the tensor, which is the linear combination of $f$-term tensors expressed in Eq. (7.1).

A three-mode tensor for the recurrence dynamics of the five mental states of an experimental phase $k$ can be mathematically expressed as

$$\underline{\mathbf{X}}_k \approx \sum_{f=1}^{F} \mathbf{a}_f^{(1)} \otimes \mathbf{a}_f^{(2)} \otimes \mathbf{a}_{f,k}^{(3)}, \tag{7.3}$$

where $\underline{\mathbf{X}}_k$ stands for the tensor of phase $k$, $k = 1, \ldots, 5$, $\mathbf{a}_f^{(1)}$ is the first-mode vector whose length is equal to the number of subject $= 1$ (single participant), $\mathbf{a}_f^{(2)}$ is the second-mode vector whose length is equal to 5 (number of mental states), and $\mathbf{a}_{f,k}^{(3)}$ is the third-mode vector whose length is equal to that of the corresponding vectorized fuzzy (joint) recurrence plot computed for phase $k$.

The solution to the above PARAFAC model is to find three factor matrices $\mathbf{a}_f^{(1)}$, $\mathbf{a}_f^{(2)}$, and $\mathbf{a}_{f,k}^{(3)}$, which can be obtained using the alternating least squares method (ALS) described in [50, 52]. The PARAFAC ALS algorithm works by successively initializing the loadings in two modes, and then iteratively estimating the last mode by the least squares regression until the estimates of the factor matrices converge.

### 7.2.3 Results

The time series of the restless, agitated, irritated, and anxious moods were used to construct the FJRP for the unrest state. The time series of the lonely and guilty moods were used to construct the FJRP for the negative state. The time series of the enthusiastic, cheerful, content, and strong moods were used to construct the FJRP for the positive state. The FRPs for the worrying and suspicious states were constructed using the time series for worrying and suspicious moods, respectively. The time points of the time series whose values were missing were removed, resulting in the time-series lengths for phases 1–5 as 176, 110, 385, 317, and 484, respectively.

The FJRPs and FRPs were constructed with the embedding dimension $= 3$, time delay $= 1$, and number of clusters $= 3$. The predefined FCM parameters for the fuzzy exponent, number of iterations, and convergence tolerance were 2, 100, and 0.00001, respectively. Based on the FJRP or FRP, the FWRN of the time series for each of the 12 moods was computed. As an illustration, Fig. 7.7 shows the time series, FRPs, and FFWRNs with a weight cutoff $< 0.5$ in order to make the graph plotting visible, for the lonely and enthusiastic moods.

Figures 7.8, 7.9, 7.10, and 7.11 show the time series of the unrest, negative, worry, suspicious, and positive states in experimental phase 2 (double blind before reducing medication) and their fuzzy joint/fuzzy recurrence plots. Figures 7.12, 7.13, 7.14, and 7.15 show the time series of the unrest, negative, worry, suspicious, and positive states in experimental phase 5 (after the experiment) and their fuzzy joint/fuzzy

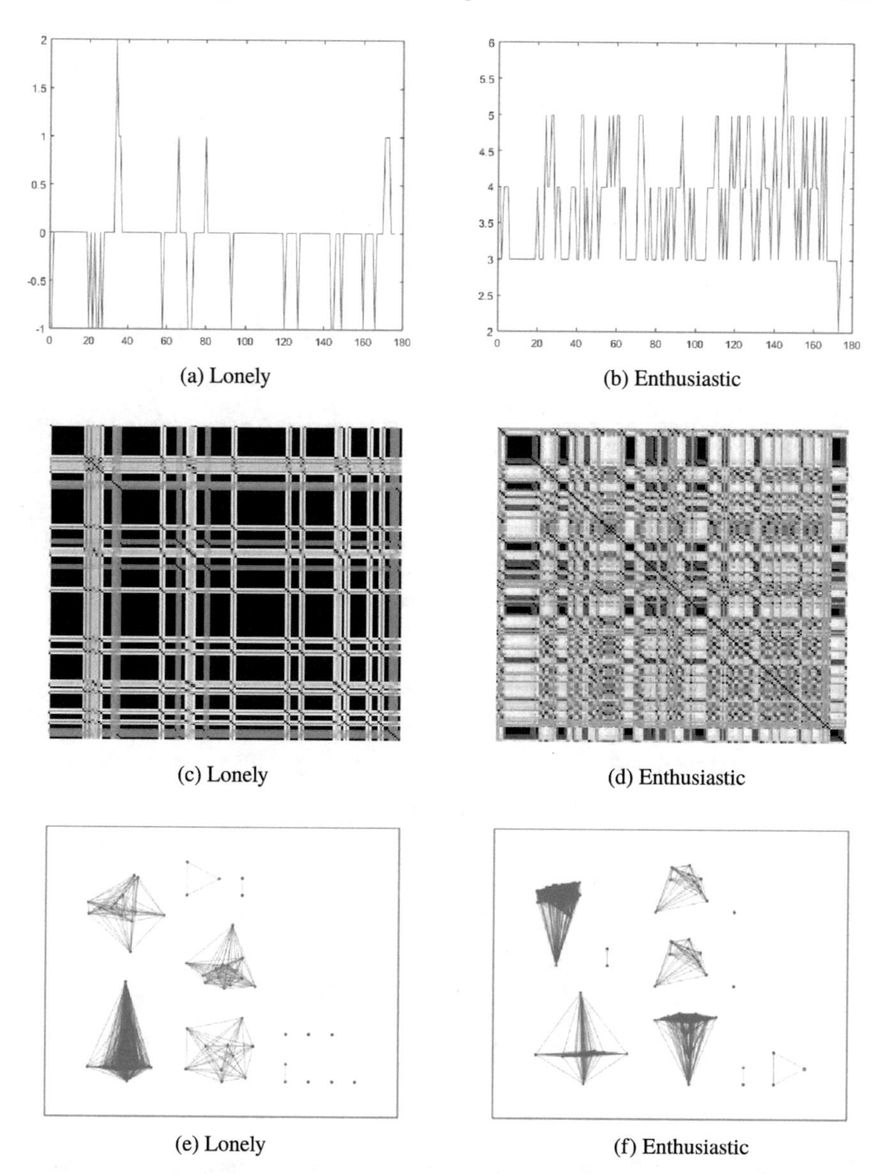


**Fig. 7.7** Time series (**a**) and (**b**), fuzzy recurrence plots (**c**) and (**d**), and fuzzy recurrence networks with cutoff $< 0.5$ (**e**) and (**f**) of lonely and enthusiastic moods in experimental phase 1 (baseline), respectively. Horizontal and vertical axes of the time series are time (days) and 7-point scale, respectively

recurrence plots. It can be seen that the degrees of recurrences of time series of the worry and suspicious states and the fuzzy joint recurrence plots of the negative states are stronger than those of the other three states in both experimental phases 2 and 5.

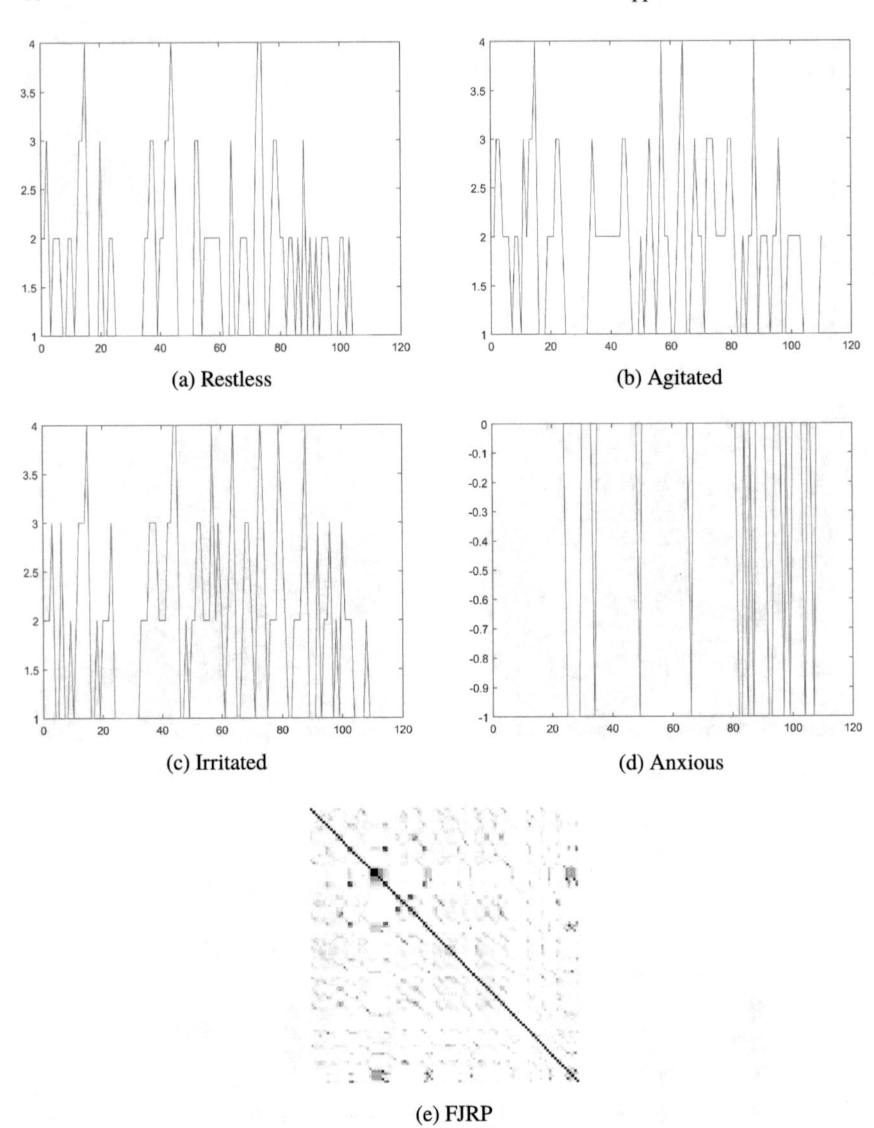


(a) Restless

(b) Agitated

(c) Irritated

(d) Anxious

(e) FJRP

**Fig. 7.8** Time series of the unrest state (**a**)–(**d**), and fuzzy joint recurrence plot of time series of the unrest state (**e**) in experimental phase 2 (double blind before reducing medication). Horizontal and vertical axes of the time series are time (days) and 7-point scale, respectively

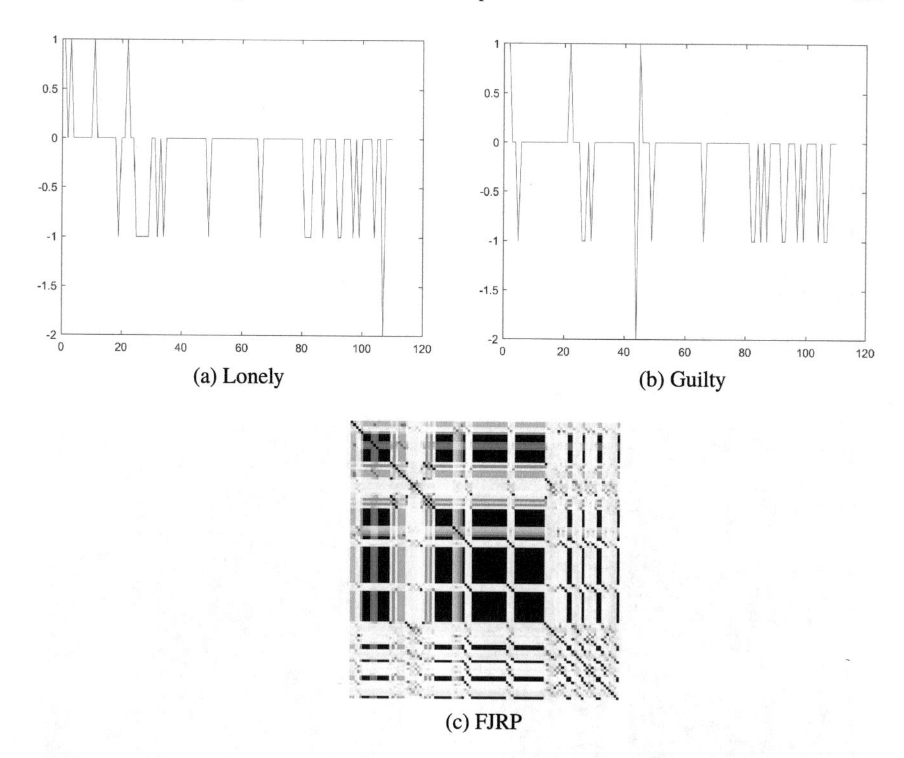


**Fig. 7.9** Time series of the negative state (**a**) and (**b**), and fuzzy joint recurrence plot of time series of the negative state (**c**) in experimental phase 2 (double blind before reducing medication). Horizontal and vertical axes of the time series are time (days) and 7-point scale, respectively

To compute the tensor decomposition for each of the five tensors, which represent five experimental phases (baseline, double blind before reducing the medication, double blind during the medication reduction, phase after the medication reduction, and phase after the experiment), and number of factors for the tensor decomposition was selected as 3.

Figure 7.16 shows the plots of the average clustering coefficient versus the characteristic path length obtained from the FWRNs of the five experimental phases for each of the five mental states, and the 3D plot of the tensor-decomposition factors for the five experimental phases for all five mental states obtained according to Eq. (7.3).

As shown in Fig. 7.16a, the subplot of the characteristic path lengths versus clustering coefficients of the five experimental phases for the unrest state shows that phase 1 (baseline) is most further away from phase 5 (phase after the experiment), while phase 2 (double blind before reducing medication) is closet to phase 1, which is intuitive, and phases 3 and 4 (phase 3 is closer to phase 1) are between the two ends. This result suggests an evolution of the unrest state after the stopping of antidepressant drug.

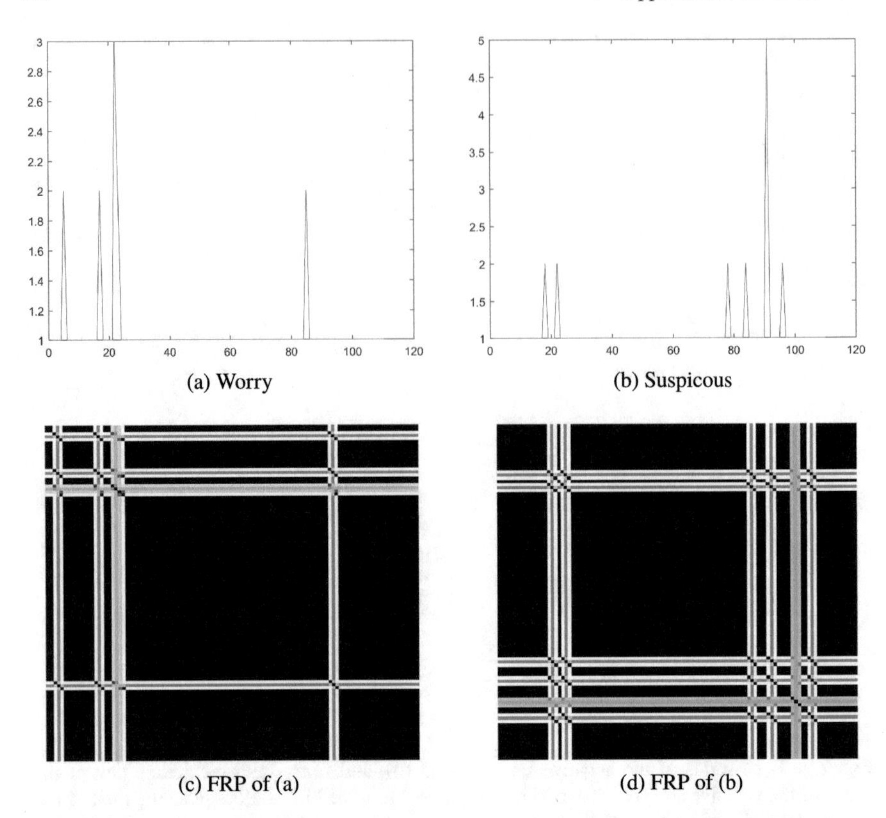


**Fig. 7.10** Time series of the worry (**a**) and suspicious state (**b**), and fuzzy recurrence plots of time series of the two states (**c**) and (**d**) in experimental phase 2 (double blind before reducing medication). Horizontal and vertical axes of the time series are time (days) and 7-point scale, respectively

For the negative affect, Fig. 7.16b shows phase 5 is closet to phase 1. The second closest point to phase 1 is phase 3, while phases 2 and 4 are the furthest two ends from phase 1. This observation indicates there was no effect of the medication on the negative affect of the participant.

Being similar to the unrest state, phase 1 is furthest away from phase 5 for the positive affect as shown in Fig. 7.16c, which also shows phases 2 and 3 are between phase 1, and phase 4 is further away from phase 1 than phase 3, indicating transition of the mental state from the gradual reduction to stopping of the medication.

For the worrying state, Fig. 7.16d shows phases 4 and 5 are further away from phase 1, but also phase 2 is far away from phase 1 in terms of the characteristic path length, while phase 3 is closest to phase 1, and phases 4 and 5 are closest to one another. In general, the plot suggests a new development of the mental state after the reduction and stopping of medication.

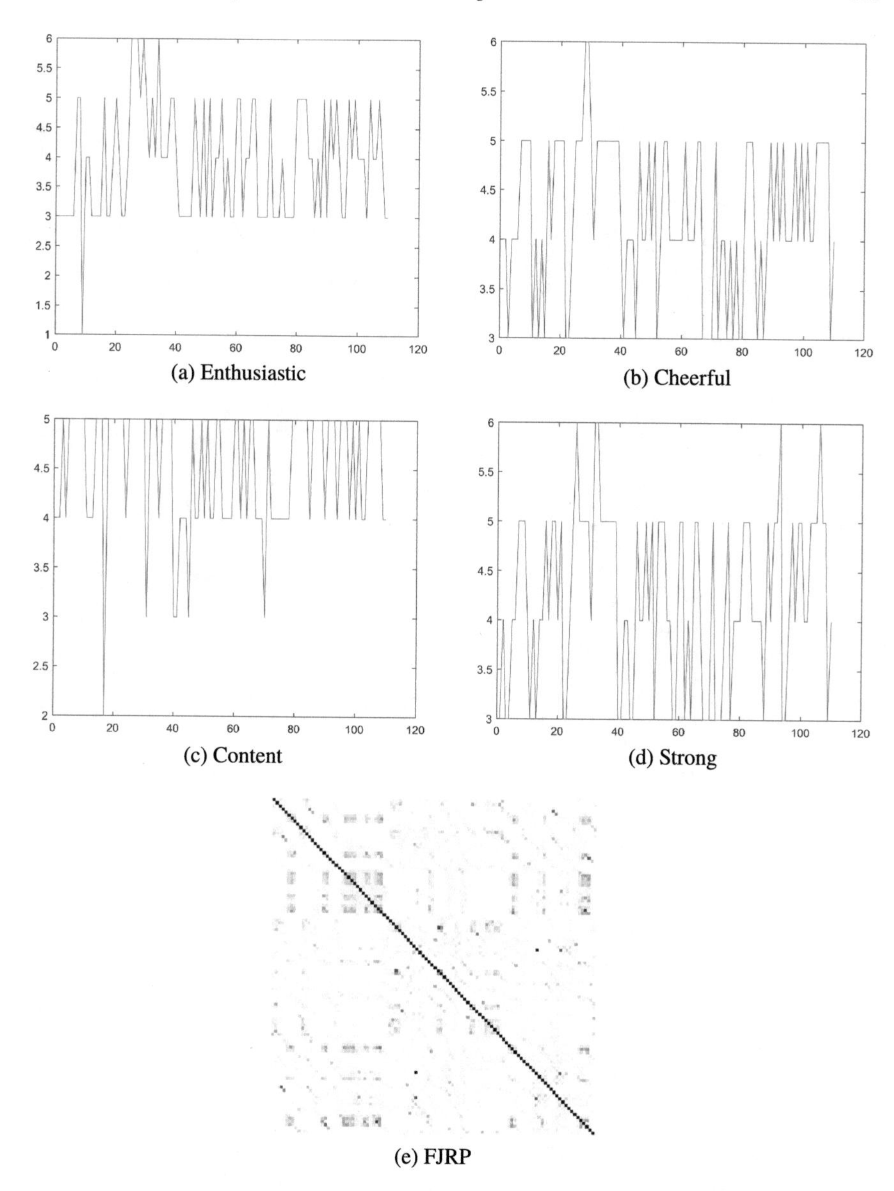


**Fig. 7.11** Time series of the positive state (**a**)–(**d**), and fuzzy joint recurrence plot of time series of the positive state (**e**) in experimental phase 2 (double blind before reducing medication). Horizontal and vertical axes of the time series are time (days) and 7-point scale, respectively

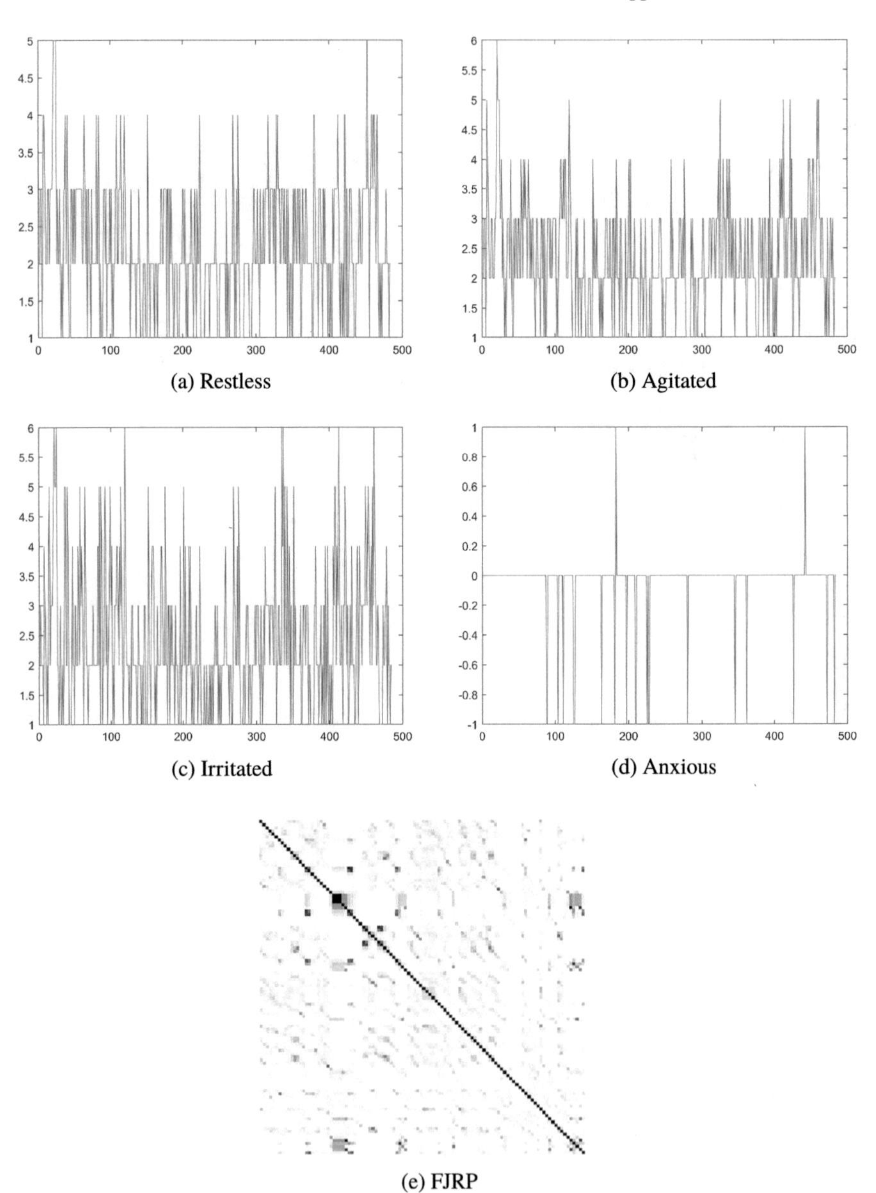


**Fig. 7.12** Time series of the unrest state (**a**)–(**d**), and fuzzy joint recurrence plot of time series of the unrest state (**e**) in experimental phase 5 (after experiment). Horizontal and vertical axes of the time series are time (days) and 7-point scale, respectively

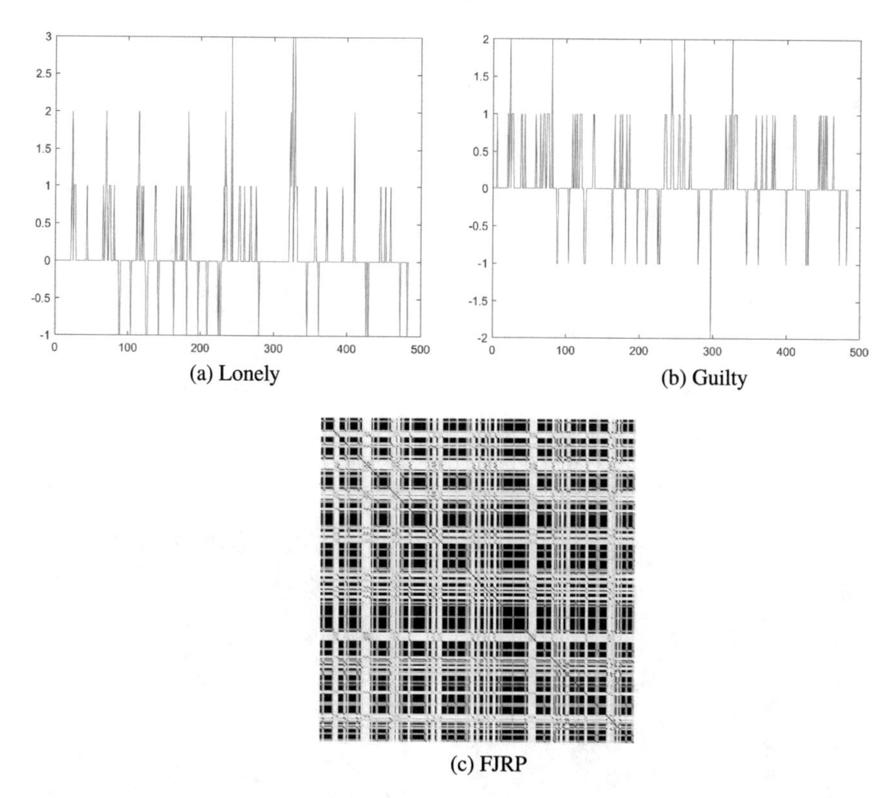


**Fig. 7.13** Time series of the negative state (**a**) and (**b**), and fuzzy joint recurrence plot of time series of the negative state (**c**) in experimental phase 2 (after experiment). Horizontal and vertical axes of the time series are time (days) and 7-point scale, respectively

Regarding the mental state of suspicion as shown in Fig. 7.16e, phase 5 is furthest from phase 1, phases 4 and 5 are closest to one another, and phases 2 and 3 are closer to phase 1. Once again, the plot supports that there is a shift in the suspicious mood after the stopping of antidepressant drug.

Figure 7.16f shows the 3D distribution of the five experimental phases by taking into account the combination of all five mental states using tensor decomposition. The spatial pattern of the phase relationship displayed in Fig. 7.16f agrees with those for the unrest, positive, worrying, and suspicious moods using the network properties. The effect of the recurrence dynamics of the negative mood was balanced by the other four mental states in the tensor decomposition.

Wichers et al. [39] showed that the participant had experienced a critical transition and the depression symptoms held to the principles of complex dynamical systems. In this study, both complex network analysis and tensor decomposition of the recurrence dynamics indicate that the participant was vulnerable to develop a new event of depression when the antidepressant medication was reduced and stopped. Such a detection in the recurrence dynamics of the data can be considered as a personalized warning signal for depression.

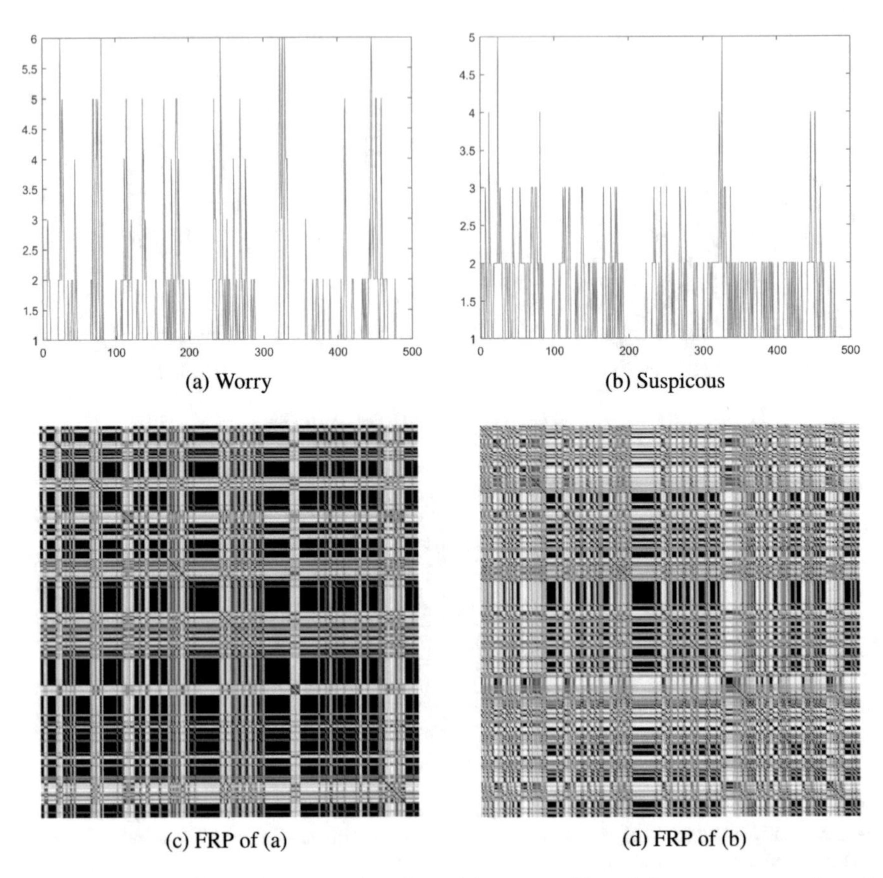


**Fig. 7.14** Time series of the worry (**a**) and suspicious state (**b**), and fuzzy recurrence plots of time series of the two states (**c**) and (**d**) in experimental phase 5 (after experiment). Horizontal and vertical axes of the time series are time (days) and 7-point scale, respectively

A new mathematical methodology for analysis of complex system dynamics of time series obtained from personalized major depression has been introduced and discussed in the foregoing sections. The results agree with the finding of a critical transition in depression of the same participant. The use of the complex network properties reveals the spatial distributions of the experimental phases with respect to different mental states, whereas the tensor decomposition of the recurrence dynamics in a multi-way fashion unveils the spatial distribution of the experimental phases with respect to all mental states. Both approaches complement each other and confirm the analysis of the depression dynamics developed over time under changes of antidepressant medication. Furthermore, the tensor decomposition presented in this study can also be useful for studying depression of different cohorts of participants by including multiple subjects in the first mode of the tensor.

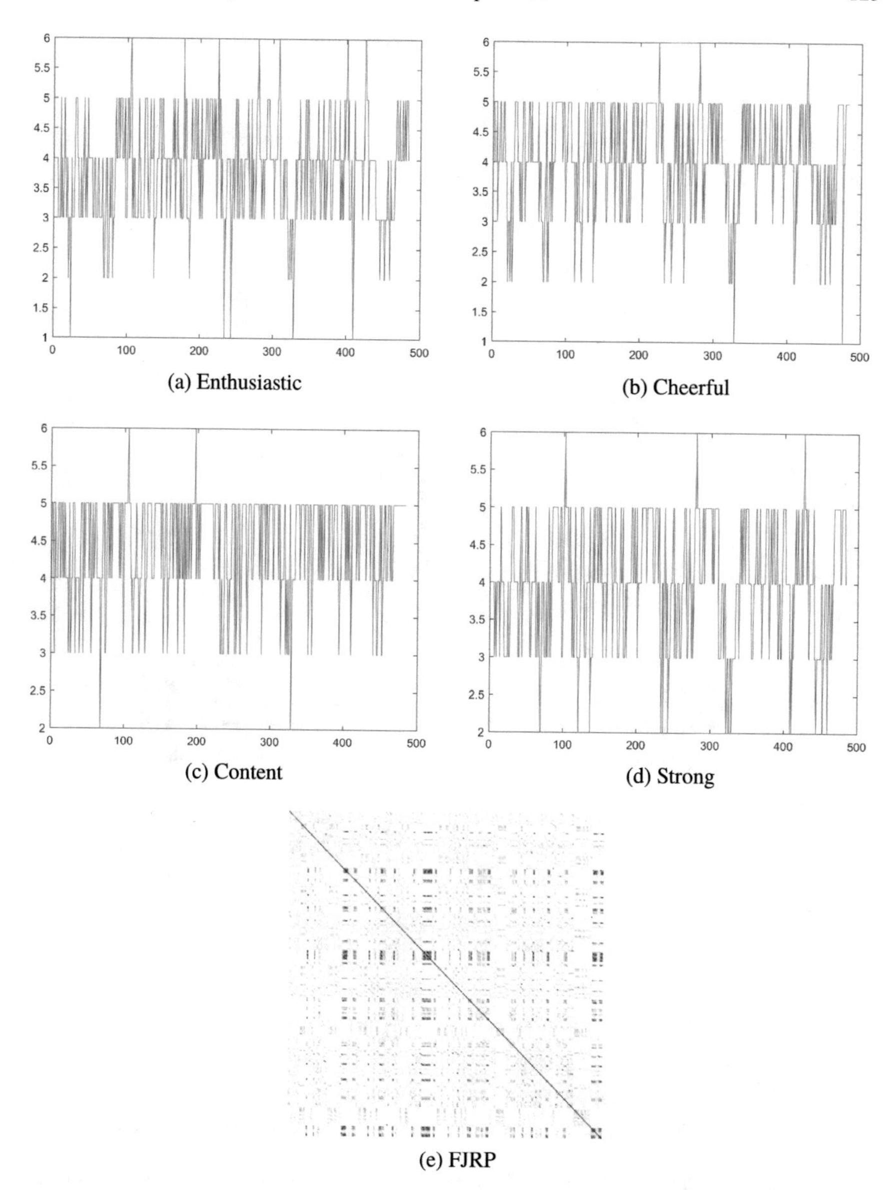


**Fig. 7.15** Time series of the positive state (**a**)–(**d**), and fuzzy joint recurrence plot of time series of the positive state (**e**) in experimental phase 5 (after experiment). Horizontal and vertical axes of the time series are time (days) and 7-point scale, respectively

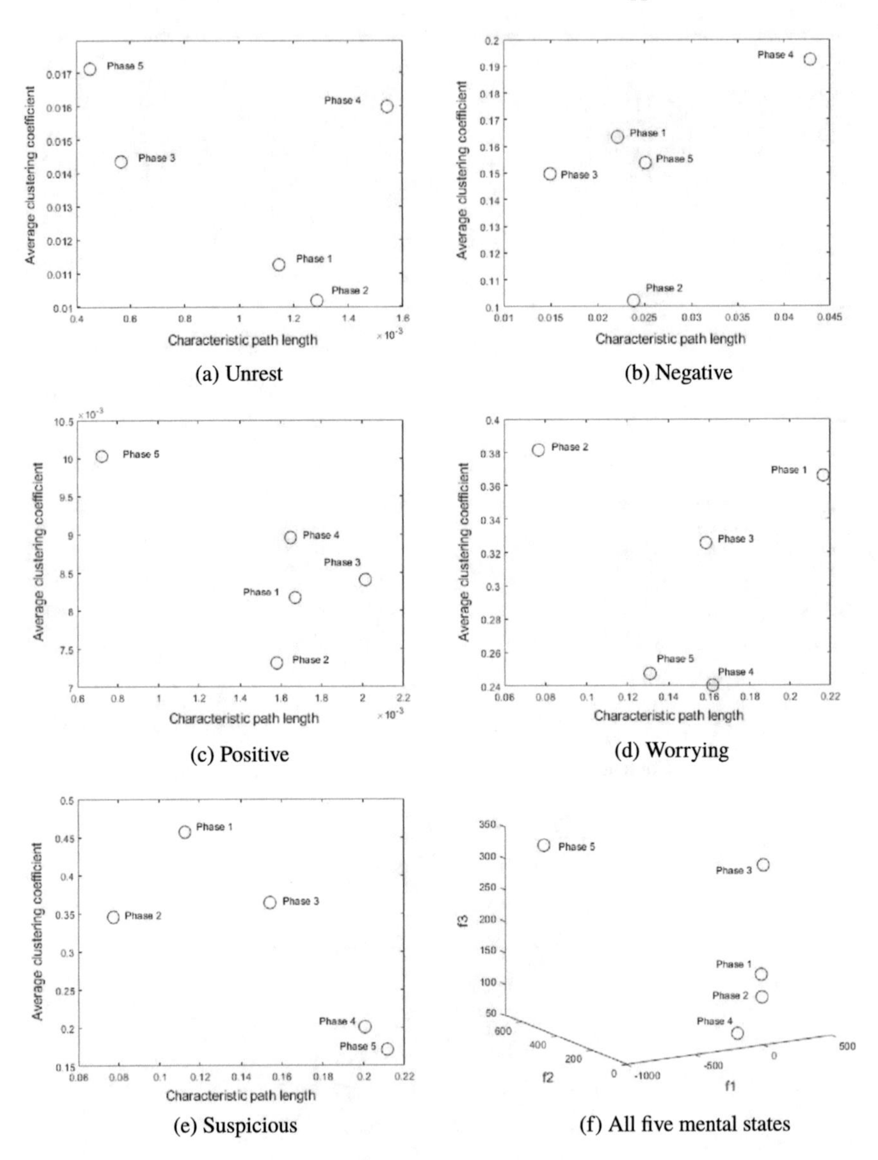


**Fig. 7.16** Plots of characteristic path length versus average clustering coefficient of individual mental states (**a**)–(**e**), and 3D plot of tensor-decomposition factors for all mental states (**f**)

## 7.3 Classification of Short Time Series with Deep Learning of Fuzzy Recurrence Plots

Parkinson's disease (PD) is a neurodegenerative disorder that affects dopaminergic neurons [53]. Statistics on PD have reported it affects approximately 10 million people worldwide, and about 4% of them before the age of 50 [54]. Symptoms of PD slowly develop over years, and the progression of PD can be different among individuals because of the diversity of the disease. People with PD can be observed with tremor, bradykinesia (slowness of movement), limb rigidity, and gait and balance problems. The cause of PD remains unknown [55]. Because a significant amount of the substantia nigra neurons have already been lost or impaired before the onset of motor features, people with PD may first start experiencing symptoms later in the course of the disease [56, 57]. Treatment options for PD can vary and include medications and sometimes deep brain stimulation [58, 59]. While Parkinson's itself is not fatal, disease complications can be serious [56].

Many scientific efforts have been spent on exploring methods for identifying biomarkers for PD [60] with the hope that these markers can lead to earlier diagnosis and targeted treatments of the disease. However, present therapies used for PD cannot slow or stop the disease in a prodromal stage [61]. There are many techniques using sensors for detecting and monitoring movement patterns on patients with PD. Most techniques using sensor-induced data focus on studying gait dynamics and temporal gait parameters [62–66]. One issue of using sensor data is that gait measurements are to be obtained during relatively long walking periods, causing discomfort to the participants or impracticability of performance in clinical settings. Therefore, research into the minimum strides required for a reliable estimation of temporal gait parameters has recently been carried out with the purpose of avoiding or minimizing discomfort to participants in gait experiments [67].

Apart from gait and balance data, the measurement of computer-keystroke time series that contain information of the hold time occurring between pressing and releasing a key collected during the sessions of typing activity using a standard word processor on a personal computer has been proposed for detecting early stages of PD [68]. Being similar to the motivation for determining the minimum number of strides for the analysis of gait dynamics, this study is interested in answering the question if there are methods that can process very short time series and achieve good results for differentiating healthy controls from subjects with early PD. If being successful, the use of computer-keystroke time series can be equivalent on a practical basis to the use of mobile sensor data for evaluating upper limb's motor functions by finger tapping [69] that is typically used in clinical trials. The finger tapping test requires a participant to press one or two buttons on a device such as an iPhone as fast as possible for a short period of time.

The method of fuzzy recurrence plots [70] developed for studying nonlinear dynamics in time series data can be useful for creating feature dimensions of short time series. Therefore, the deep learning of fuzzy recurrence plots was proposed

[71], for the classification of short time series of computer-key hold time recorded from two cohorts of healthy control and early PD.

Having highlighted earlier, the concept of using fuzzy recurrence plots of short time series for classification using long short-term memory (LSTM) networks is original in that it contributes to the increase of the feature dimensions of short raw time series, which, in turn, can improve the classification. No prior work of similar concepts exists in literature. A survey of recent reports that applied LSTM and convolutional neural networks for time series or sequential data classification is addressed herein.

Regarding time-series classification methods, the deep-learning method of LSTM neural networks is currently known as a state-of-the-art model for the classification of time series or sequential data [72], including speech recognition [73] and machine translation [74, 75]. A deep recurrent neural network called TimeNet for extracting features from time series was developed [76]. The TimeNet was designed to generalize time series representation across domains. A trained TimeNet can be used as a feature extractor for time series and was reported to be useful for time series classification by performing better than a domain-specific recurrent neural network and dynamic time warping [76]. Stacked LSTM autoencoder networks were applied to extract features of time-series data, which were then used to train deep feedforward neural networks for classification of multivariate time series recorded with sensors in the steel industry to detect steel surface defects [77]. In this work, the features extracted with LSTM autoencoders were found to be useful, and therefore the need for domain expert knowledge can be alleviated.

Other time-series classification using convolutional neural networks (CNNs) have recently been introduced. A convolutional LSTM (ConvLSTM) was introduced for a spatiotemporal sequence forecasting problem in which both the input and the prediction target are spatiotemporal sequences [78]. This ConvLSTM model was constructed by extending the fully connected LSTM to have convolutional structures in the input-to-state and state-to-state transitions. The ConvLSTM network was reported to perform better than the fully connected LSTM by being able to capture the spatiotemporal correlations of the sequential data for precipitation nowcasting. A multiscale convolutional neural network [79], which extracts deep-learning features at different scales and frequencies from three representations of time series including the original, down-sampled, and smoothed data, was reported to be capable of extracting effective features for time-series classification. Baseline full convolutional networks were proposed for time series classification [80]. The proposed baseline models were reported to be pure end-to-end without demanding heavy preprocessing of the raw data or feature crafting, and achieve competitive performance to other state-of-the-art approaches, including the multilayer perceptron, fully convolutional network, and residual network. LSTM fully convolutional network models were introduced for time-series classification [81]. This network consists of a branching structure. The first branch is the convolutional part, whereas the second branch is an LSTM block which receives a time series in a transposed form as multivariate time series with a single time step. The outputs of the two branches are concatenated and then fed to a classifier. These models were reported to be able to enhance the performance of fully convolutional networks with a nominal increase in model size as well as require minimal data preprocessing.

In general, time-series classification has been recognized as an important and challenging area of research, particularly with respect to the demand for handling increasing availability of new data of time series. While numerous algorithms for time-series classification have been published in literature and the popularity of deep learning has been pervasive in many disciplines, only a few deep neural networks have been applied to solving time-series classification problems. A recent survey on promising applications of deep neural networks for time series classification in several areas can be found in [82]. This study introduces the usefulness of constructing fuzzy recurrence plots of short time series that can be incorporated into LSTM models to improve the classification accuracy.

### *7.3.1 LSTM Neural Networks with FRPs*

An LSTM neural network [83] is an artificial recurrent neural network (RNN) used in deep learning. Unlike conventional feedforward neural networks, an LSTM model has feedback loops that allow information of previous events to be carried on in the sequential learning process. Therefore, LSTM networks are effective in learning and classifying sequential data such as speech and video analysis [84–87].

The internal state of an RNN is used as a memory cell to map real values of input sequences to those of output sequences that reflect the dynamic pattern of time series, and therefore is considered an effective algorithm for learning and modeling temporal data [88]. However, because an RNN uses sequential processing over time steps that can easily degrade the parameters capturing short-term dependencies of information sequentially passing through all cells before arriving at the current processing cell. This effect causes the gradient of the output error with respect to previous inputs to vanish by the multiplication of many small numbers being less than zero. This problem is known as vanishing gradients [89]. LSTM networks attempt to overcome the problem of the vanishing gradients encountered by conventional RNNs by using gates to keep relevant information and forget irrelevant information.

The difference between an LSTM neural network and a conventional RNN is the use of memory blocks for the former network instead of hidden units for the latter [90]. The input gate of an LSTM network guides the input activations into the memory cell and the output gate carries out the output flow of cell activations into the rest of the network. The forget gate allows the flow of information from the memory block to the cell as an additive input, therefore adaptively forgetting or resetting the cell memory. Thus, being less sensitive to the time steps makes LSTM networks better for analysis of sequential data than conventional RNNs. A common LSTM model is composed of a memory cell, an input gate, an output gate, and a forget gate. The cell memorizes values over time steps and the three gates control the flow of information into and out of the cell. The weights and biases to the input gate regulate the amount of new value flowing into the cell, while the weights and

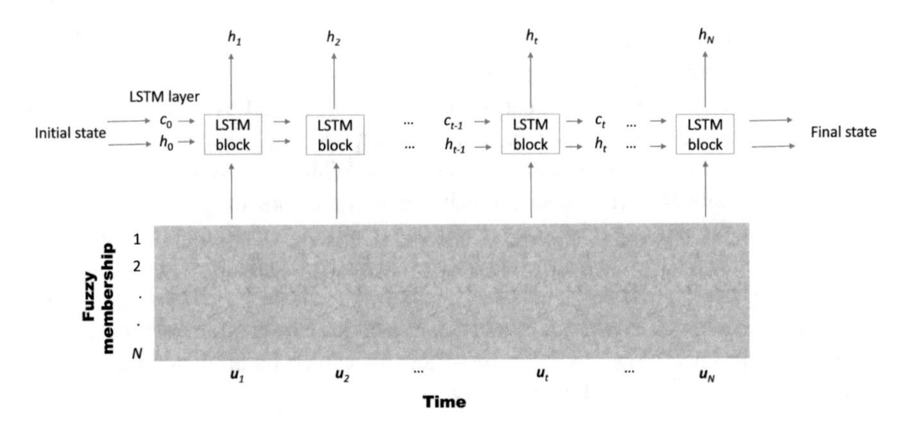


**Fig. 7.17** For an LSTM layer, the first LSTM block takes the initial state of the network and the first time step of the FRP, and then computes the first output $h_1$ and the updated cell state $c_1$, then at time step $t$, the LSTM block takes the current state of the network $c_{t-1}, h_{t-1}$ and the next time step of the FRP at $t$, and then computes the output $h_t$ and the updated cell state $c_t$. Note: Fuzzy membership and time steps of FRP are not drawn to scale, and the image gradient is a virtual representation of the fuzzy membership of recurrences

biases to the forget gate and output gate control the amount of information to remain in the cell and the extent to which the value in the cell is used to compute the output activation of the LSTM block, respectively.

The architecture for an LSTM block, in which the fuzzy membership grades of the FRP are the input values, is described in Fig. 7.17. Furthermore, it can be visualized from Fig. 7.17 that the use of FRPs can increase the feature of a time series from one dimension to $N$ dimensions, where $N$ is the number of the phase-space vectors of the time series, enhancing the training of the LSTM network. The mathematical expressions for the four gates of an LSTM block at time step $t$ with the input of a FRP represented with its discrete fuzzy membership vector of $N$ dimensions at time $t$, denoted as $\mathbf{u}_t = (\mu_t^1, \mu_t^2, \ldots, \mu_t^N)^T$, are given as follows [83]:

$$\mathbf{f}_t = \sigma_g(\mathbf{W}_f\mathbf{u}_t + \mathbf{R}_f\mathbf{h}_{t-1} + \mathbf{b}_f), \tag{7.4}$$

where $\mathbf{f}_t \in \mathbb{R}^N$ is the activation vector of the forget gate at time $t$, $\sigma_g$ denotes the sigmoid function, $\mathbf{W}_f \in \mathbb{R}^{M\times N}$ is the input weight matrix, $M$ refers to the number of hidden layers, $\mathbf{R}_f \in \mathbb{R}^{M\times M}$ is the recurrent weight matrix, $\mathbf{h}_{t-1} \in \mathbb{R}^M$ is the hidden state vector at time $(t-1)$, which is also known as the output vector of the LSTM unit and the initial $\mathbf{h}_0 = 0$, and $\mathbf{b}_f \in \mathbb{R}^M$ is the bias vector of the forget gate.

The input gate at time $t$, denoted as $\mathbf{i}_t \in \mathbb{R}^M$, is expressed as

$$\mathbf{i}_t = \sigma_g(\mathbf{W}_i\mathbf{u}_t + \mathbf{R}_i\mathbf{h}_{t-1} + \mathbf{b}_i), \tag{7.5}$$

where $\mathbf{W}_i$, $\mathbf{R}_i$, and $\mathbf{b}_i$ are similarly defined as in Eq. (7.4).

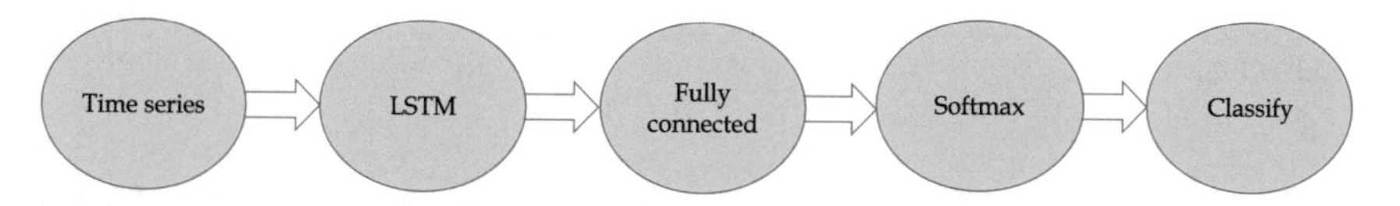


**Fig. 7.18** Architecture of an LSTM network for time-series classification

The cell candidate vector that adds information to the cell state at time step $t$, denoted as $\mathbf{g}_t \in \mathbb{R}^M$, is defined as

$$\mathbf{g}_t = \sigma_c(\mathbf{W}_g \mathbf{u}_t + \mathbf{R}_g \mathbf{h}_{t-1} + \mathbf{b}_g), \tag{7.6}$$

where $\sigma_c$ is hyperbolic tangent function (tanh), $\mathbf{W}_g$, $\mathbf{R}_g$, and $\mathbf{b}_g$ are similarly defined as in Eq. (7.4).

The output gate, denoted as $\mathbf{o}_t \in \mathbb{R}^M$, which controls the level of the cell state added to hidden state is expressed as

$$\mathbf{o}_t = \sigma_g(\mathbf{W}_o \mathbf{u}_t + \mathbf{R}_o \mathbf{h}_{t-1} + \mathbf{b}_o), \tag{7.7}$$

where $\mathbf{W}_o$, $\mathbf{R}_o$, and $\mathbf{b}_o$ are similarly defined as in Eq. (7.4).

The cell state vector at time step $t$, denoted as $\mathbf{c}_t \in \mathbb{R}^M$, is given by

$$\mathbf{c}_t = \mathbf{f}_t \circ \mathbf{c}_{t-1} + \mathbf{i}_t \circ \mathbf{g}_t, \tag{7.8}$$

where the initial values for $\mathbf{c}_0 = 0$, and the operator $\circ$ denotes the Hadamard product.

Finally, the hidden state vector at time step $t$ is given by

$$\mathbf{h}_t = \mathbf{o}_t \circ \sigma_c(\mathbf{c}_t). \tag{7.9}$$

Figure 7.18 illustrates the architecture of an LSTM network for classification of time series. The network starts with an input layer of time series followed by an LSTM layer. To predict class labels, the LSTM network ends with a fully connected layer, a softmax layer, and a classification output layer.

### *7.3.2 Database*

The neuroQWERTY MIT-CSXPD database [91], which is publicly available from the PhysioNet (research resource for complex physiologic complex signals), was used in this study. The data contains keystroke logs collected from 85 subjects with and without PD. This dataset was collected and analyzed for investigating if the routine interaction with computer keyboards can be used to detect motor signs in the early stages of the PD subjects whose average time since diagnosis was 3.9 years, who were on PD medication but had no medication for the 18 hours before the typing test [68]. The subjects were recruited from two movement disorder units in Madrid, Spain, following the institutional protocols approved by the Massachusetts Institute of Technology, USA, Hospital 12 de Octubre, Spain, and Hospital Clinico San Carlos, Spain.

Each data file collected includes the timing information collected during the sessions of typing activity using a standard word processor on a Lenovo G50-70 i3-4005U with 4MB of memory and a 15 inches screen running Manjaro Linux. The lengths of computer-key hold time series are highly variable, some have around 500, and others around 2500 time points.

Subjects were instructed to type as they normally would do at home and they were left free to correct typing mistakes only if they wanted to. The key acquisition software presented a temporal resolution of 3/0.28 (mean/standard deviation) milliseconds. Along with the raw typing collections, clinical evaluations were also performed on each subject, including UPDRS-III (Unified Parkinson's Disease Rating Scale: Part III) [92] and finger tapping tests.

### *7.3.3 Results*

Based on a previous study [93], outliers existing in the raw time series of the HC and PD individuals, where several data points are in the magnitude of $10^9$, were removed from the time series. Because the purpose of this study is to classify the signals of short lengths, short segments from the start of the original time series were selected for testing the use of the LSTM neural-network model with FRPs.

Figure 7.19 shows two short time series of 50 time steps of the computer-key hold durations, extracted from the start of the original time series, recorded from a healthy control (HC) and early PD subjects, and their associated RPs and FRPs with an embedding dimension $= 3$, and time delay $= 1$, similarity tolerance $= 0.1$ for RPs, and number of clusters $= 5$ for FRPs. It can be observed from Fig. 7.19 that the binary information obtained from the RPs are very sparse for both HC and early PD subjects, lacking useful feature information. The RPs become sparser with increased embedding dimensions of 3 and 5. The FRPs display rich information as texture images as the values of the fuzzy membership grades about the recurrences of the underlying dynamics of the time series of the two subjects. Therefore, the fuzzy membership grades of the phase-space vectors of the time series were used as the feature with dimensions being equal to the number of the phase-space vectors for training and classification using the LSTM network.

The LSTM neural network of the Matlab Deep Learning Toolbox (R2018b) was used in this study. The number of hidden layers $= 100$, maximum number of epochs $= 200$, and learning rate $= 0.001$. $L_2$ regularization was used for the biases, input weights, and recurrent weights to reduce model overfitting. To construct FRPs, FCM parameters that are the fuzzy weighting exponent $w$, the number of clusters $c$, and the maximum number of iterations were chosen to be 2, 6, and 100, respectively. Given

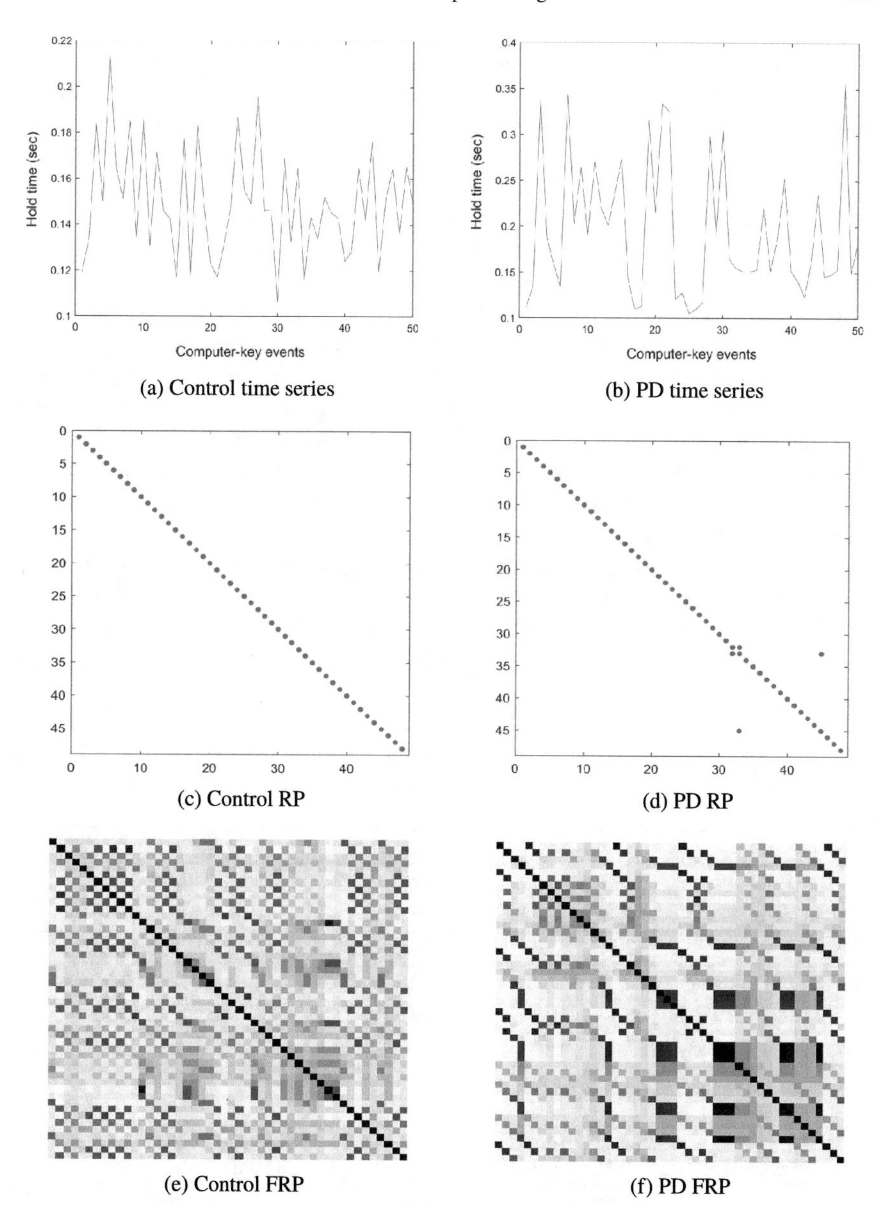


(a) Control time series (b) PD time series

(c) Control RP (d) PD RP

(e) Control FRP (f) PD FRP

**Fig. 7.19** Time series, and corresponding RPs, and FRPs of a control subject and an early PD subject. Note: RPs are displayed as the sparsity patterns of the RP matrices to make the plots visualizable, while FRPs are shown as grayscale images of the FRP matrices an embedding dimension $m$, time delay $\tau$, the number of the phase-space vectors of a time series of length $L$, which is also the number of feature dimensions used in the LSTM network, is calculated as $N = L - (m - 1)\tau$. Thus, keeping $\tau = 1$, the feature dimensions for $m = 1, 3$, and 5 for L = 50 are 50, 48, and 46, respectively, and for L = 30, the feature dimensions for $m = 1, 3$, and 5 are 30, 28, and 26, respectively. The time delay was set to be 1 for both lengths of the time series. There are methods for estimating the time delay and embedding dimension for the phase-space reconstruction such as the false nearest neighbor (FNN) and average mutual information (AMI), respectively, where the first local minima of the FNN and AMI functions are indicative of the embedding dimension and time delay, respectively [94]. However, estimate for the embedding dimension and time delay for the phase-space reconstruction of each time series of the computer-key hold duration is not convenient for implementing the LSTM network because of the variation in the feature dimensions. It has been reported that the use of time delay = 1 is well adopted for studying nonlinear time series [93], and several embedding dimensions were adopted in this study.

Tables 7.4 and 7.5 show the results of classifying HC and early PD subjects using the short time series of lengths 50 and 30, respectively. Values of the accuracy, sensitivity, and specificity are based on the average of five repetitions of the tenfold cross-validation results. The sensitivity, which is also called the true positive rate,

**Table 7.4** Average accuracy (%), sensitivity (%), and specificity (%) rates obtained from classification of control and early PD using short time series of length = 50 and different methods

| Method | Accuracy | Sensitivity | Specificity |
|---|---|---|---|
| LSTM-Time series | $63.43 \pm 4.55$ | $100 \pm 0.00$ | $0.00 \pm 0.00$ |
| CNN-GoogLeNet | $54.29 \pm 18.63$ | $65.00 \pm 28.50$ | $40.00 \pm 27.89$ |
| CNN-AlexNet | $37.14 \pm 7.82$ | $35.00 \pm 28.50$ | $40.00 \pm 43.46$ |
| LSTM-FRP ($m = 1$) | $72.00 \pm 15.92$ | $90.00 \pm 22.36$ | $46.67 \pm 50.55$ |
| LSTM-FRP ($m = 3$) | $65.14 \pm 11.50$ | $66.67 \pm 33.33$ | $63.33 \pm 41.50$ |
| LSTM-FRP ($m = 5$) | $63.43 \pm 4.55$ | $100 \pm 0.00$ | $0.00 \pm 0.00$ |

**Table 7.5** Average accuracy (%), sensitivity (%), and specificity (%) rates obtained from classification of control and early PD using short time series of length = 30 and different methods

| Method | Accuracy | Sensitivity | Specificity |
|---|---|---|---|
| LSTM-Time series | $62.10 \pm 4.33$ | $93.33 \pm 14.91$ | $10.00 \pm 22.36$ |
| CNN-GoogLeNet | $65.71 \pm 21.67$ | $70.00 \pm 20.92$ | $60.00 \pm 27.89$ |
| CNN-AlexNet | $68.57 \pm 6.39$ | $55.00 \pm 11.18$ | $86.67 \pm 29.81$ |
| LSTM-FRP ($m = 1$) | $72.38 \pm 11.24$ | $78.33 \pm 21.73$ | $66.67 \pm 23.57$ |
| LSTM-FRP ($m = 3$) | $81.90 \pm 11.74$ | $95.00 \pm 11.18$ | $66.67 \pm 23.57$ |
| LSTM-FRP ($m = 5$) | $70.10 \pm 9.63$ | $95.00 \pm 11.18$ | $36.67 \pm 24.72$ |

measures the proportion of actual positives (early PD subjects) that are correctly identified as such, whereas specificity, which is also known as the true negative rate, measures the proportion of actual negatives (control subjects) that are correctly identified as such. For the direct use of the time series (LSTM-Time series) of length = 50, accuracy = 63%, sensitivity = 100%, and specificity = 0%. For the use of FRPs of the time series (LSTM-FRP) of length = 50, with embedding dimension $m = 1$, accuracy = 72%, sensitivity = 90%, and specificity = 47%; with $m = 3$, accuracy = 65%, sensitivity = 67%, and specificity = 63%; and for $m = 5$, accuracy = 63%, sensitivity = 100%, and specificity = 0%. The accuracy results obtained from the use of FRPs are equal to or higher than the accuracy obtained from the direct use of the time series. With the use of FRPs as features, the accuracy decreases with increasing value for $m$, where the best accuracy is obtained with $m = 1$. The direct use of the time series gives 100% for sensitivity but 0% for sensitivity, and such results are not practically helpful because all are identified as early PD, which may lead to a large false positive rate.

For the time series of shorter length = 30, the direct use of the time series for LSTM network training and validation results in accuracy = 62%, sensitivity = 93%, and specificity = 10%. Once again, while the sensitivity is very high, the specificity is very low. The results obtained from the direct use of the time series for both time series of lengths = 50 and 30 are similar in accuracy, sensitivity, and specificity. For the use of FRPs of the time series of length = 30, with $m = 1$, accuracy = 72%, sensitivity = 78%, and specificity = 67%; with $m = 3$, accuracy = 82%, sensitivity = 95%, and specificity = 67%; and for $m = 5$, accuracy = 70%, sensitivity = 95%, and specificity = 37%. The FRPs with $m = 3$ give the highest accuracy (82%) among the others. All accuracy results obtained from the FRPs are higher than those obtained from the direct use of the time series.

The standard deviations of the results obtained from the use of FRPs for signal length = 50 with $m = 1$ and 3 are higher than those of the raw time series, because some rates of accuracy obtained from the FRPs reached 100% and 85%, respectively. However, even the average accuracy obtained from the use of the raw signals of the same length is lower than those obtained from the FRPs, and the average specificity (true negative rate that is the ability to correctly identify those without PD) obtained from using the raw signals is zero, which is obviously not useful at all (Table 7.4). Similarly, the standard deviations of the accuracy results obtained from the use of FRPs for signal length = 50 with $m = 1$, 3, and 5 are higher than those of the raw time series, because some rates of accuracy obtained from the FRPs reached 83%, 100%, and 86%, respectively. Once again, even the average accuracy obtained from the use of the raw signals of the same length is lower than those obtained from the FRPs, and the average specificity obtained from using the raw signals is very low (10%), which is not useful for the classification (Table 7.5).

As an illustration for the performance of the FRPs preferred to that of the direct use of the short time series, regarding the LSTM training, each iteration is an estimation of the gradient and an update of the network parameters. For the time-series length of 30, the accuracy of the direct use of the short time series converged to around 60% and the loss around 0.7, while the accuracy and loss for the use of the FRPs converged to 100% and 0, respectively. Furthermore, for the direct use of the time series in the LSTM network, the longer time series ($L = 50$) yields higher accuracy than the shorter ones ($L = 30$), but for the use of the FRPs, the accuracy depends on the selection of the embedding dimension ($m$) parameter, suggesting the influence of the embedding dimension over the time-series length, and potential exploration of the use of FRPs for classifying short sequences. The accuracy values obtained from the direct input of the time series of two different lengths are similar, while these are highly variable for the FRPs. The augmentation of more training data for the two classes would be expected to reduce the accuracy variation obtained from the input of the FRPs. Another potential factor for the higher accuracy obtained from the FRPs of the shorter time series is the redundancy with respect to the higher feature dimensions provided by the FRPs of the longer time series. This factor also suggests the ability of FRPs to extract effective dynamical features from short time series with an appropriate selection of collective parameters for the phase-space reconstruction of different time series.

Tables 7.4 and 7.5 also show the average cross-validation results of classifying HC and early PD subjects obtained from two popular pretrained deep CNNs known as GoogLeNet [95] and AlexNet [96], using the short time series of lengths 50 and 30, respectively. The implementations of these two pretrained CNN models for time-series classification were based on the work proposed in [97]. It is known that training a deep CNN from scratch is computationally expensive and requires a large amount of training data. In this study, a large amount of training data is not available. Thus, taking advantage of existing deep CNNs that have been trained on large datasets for conceptually similar tasks is desirable. This leveraging of existing neural networks is called transfer learning. GoogLeNet and AlexNet, which were pretrained for image recognition, were adopted to classify transformed images of the short time series based on a time-frequency representation [97]. Scalograms were used to obtain the RGB images of time-frequency representations of the time series. A scalogram is the absolute value of the continuous wavelet transform (CWT) coefficients of a signal. Parameters used for obtaining the scalograms of the time series and modifying GoogLeNet and AlexNet for the time-series classification are the same as described in [97]. For the time series of length = 50, classification results obtained from both deep convolutional networks, GoogLeNet (CNN-GoogLeNet) and AlexNet (CNN-AlexNet), are lower than LSTM-Time series and LSTM-FRP. CNN-AlexNet has the lowest average accuracy (37%). For the time series of shorter length = 30, classification results obtained from CNN-AlexNet (69%) are higher than those obtained from the CNN-GoogLeNet and LSTM-Time series, but lower than the LSTM-FRP.

Once again, the idea of using FRPs of short raw time series for classification with an LSTM network is original, and there are no previous reports on this kind of research. Furthermore, on the comparison between the LSTM-based classification using raw time series and FRPs of raw time series, an appropriate construction of an FRP in order to correctly reflect the dynamics underlying the signal is mainly subject to the selection of a good embedding dimension $m$, whereas $m$ being not applicable for the case of LSTM-based classification of raw time series. Hence, three different values of $m$ were chosen for the construction of FRPs of the raw signals to gain insight into the influence of the embedding dimension over the classification. The LSTM-based classification using any of the three values for $m$ (except with $m = 5$ for length $= 50$, the accuracy rates of LSTM-FRP and LSTM-Time series are the same) specified for constructing the FRPs outperformed those using LSTM with raw time series and the two pretrained CNN models.

This study attempts to show that the incorporation of FRPs of short time series, which creates several dimensions or channels for each time step of the time series, as input into the LSTM model can improve the LSTM-based classification. The number of dimensions associated with the time steps of a sequence is considered as the number of features flown through an LSTM layer, which constitutes to the LSTM layer architecture as described by LSTM networks in the Matlab Deep Learning Toolbox (R2018b and R2019a). The core components of an LSTM network are a time-series input layer and an LSTM layer. The input layer inputs time series into the network. An LSTM layer learns long-term dependencies between time steps of the data.

The purpose of classifying short time series is to reduce potential discomfort caused to individuals participating in the test. The use of the fuzzy membership grades of the recurrences of the phase-space vectors of the time series contributes to the increase of the feature dimensions for the learning of the LSTM network, and therefore can improve the classification over the direct input of the time series. The results obtained from the FRPs are encouraging for the collection of practical data recorded from participants and their usage for the classification task.

The selection of a value for the embedding dimension of the phase-space reconstruction of the time series can influence the classification. Designing an optimal procedure for estimating collective embedding dimensions as well as time delays for the sets of time series of different classes would be worth investigating to improve the effective use of FRPs in LSTM networks for short time-series classification. Furthermore, application of bidirectional LSTM (BiLSTM) networks [84] that is based on the concept of bidirectional RNNs [98] to simultaneously learn from past (backward) and future (forward) dependencies between time steps of time series or sequence data is worth exploring for improving the classification.

In this study, the FRPs of short time series were used as sequential data for classification. The inherent texture of FRPs can be extracted with several methods of texture analysis in image processing [66] or pretrained image-based deep-learning models for classification as a transformation of one-dimensional signals into two-dimensional images. Furthermore, the approach proposed in this study is not limited to differentiating healthy control from early PD subjects with short time series but can also be applied to other problems concerning with machine learning and classification using short time-series data.

## 7.4 Analysis of Computer-Keystroke Time Series

This study [93] used the same neuroQWERTY MIT-CSXPD database described in Sect. 7.3 for the analysis of computer-keystroke time series obtained from healthy control (HC) and early PD subjects with RPs and FRPs. Due to outliers existing in the raw time series of the HC and PD individuals, where several data points are in the magnitude of $10^9$, those huge values were removed from the time series before carrying out the task of feature extraction. To extract RQA features, the RPs of the HC and PD individuals were first computed with embedding dimension $= 1$, time delay $= 1$, and similarity threshold $= 0.05$ of the mean of the time series.

Eleven RQA features extracted from the RPs are recurrence rate (*RR*), determinism (*DET*), mean diagonal line length ($< L >$), maximal diagonal line length ($L_{\max}$), entropy of the diagonal line lengths (*ENT*), laminarity (*LAM*), trapping time (*TT*), maximal vertical line length ($V_{\max}$), recurrence time of the first type ($T_1$), recurrence time of the second type ($T_2$), and recurrence time entropy (*RTE*). Both minimal diagonal line length $L_{\min}$ and minimal vertical line length $V_{\min}$ were set to 2.

To extract FRP-based GLCM features, the FRPs of the HC and PD individuals were first computed with embedding dimension $= 1$, time delay $= 1$, and the number of clusters $= 5$. Nineteen GLCM-based features were extracted from the FRPs, which are autocorrelation, cluster prominence, cluster shade, contrast, correlation, difference entropy, difference variance, dissimilarity, energy, entropy, homogeneity, information measure of correlation 1, information measure of correlation 2, inverse difference, maximum probability, sum average, sum entropy, sum of squares variance, and sum variance. The orientation for computing the GLCM is one pixel to the right.

To compute various values of the $< CC >$ (the average clustering coefficient) and *CPL* (the characteristic path length), the scalable recurrence networks of the HC and PD individuals were constructed with the number of clusters $= 20$, $\alpha = 0.5$, and $\beta = 0.01, 0.05$, and $0.1$.

The means and standard deviations of 11 RQA features for the HC and PD groups are shown in Table 7.7. The means and standard deviations of 19 FRP-based GLCM features for the HC and PD groups are shown in Table 7.6. Pearson's correlation coefficient for the 11 RQA features of the HC and PD is 1 with the corresponding $p$-value of zero to the 20 digits, whereas Pearson's correlation coefficient for the 19 GLCM features of the HC and PD is 0.9960 with the corresponding $p$-value of zero to the 18 digits. Such values indicate a very strong correlation between the HC and PD features and suggest better discriminating power of the GLCM features extracted from fuzzy recurrence plots.

The twofold cross-validations of the classification of the HC subjects and PD patients using the least squares support vector machines (LS-SVM) [99] were repeated 10 times. The mean values of the receiver operating characteristics (ROC) obtained from the RQA and FRP-GLCM features, as well as the areas under the ROC curves (AUC) obtained from the numerical neuroQWERTY index (nQi) model [68], where nQi1 is the model using all nonoverlapping windows of the time series,

**Table 7.6** Means and standard deviations of 19 FRP-based GLCM features, where F1 = autocorrelation, F2 = cluster prominence, F3 = cluster shade, F4 = contrast, F5 = correlation, F6 = difference entropy, F7 = difference variance, F8 = dissimilarity, F9 = energy, F10 = entropy, F11 = homogeneity, F12 = information measure of correlation 1, F13 = information measure of correlation 2, F14 = inverse difference, F15 = maximum probability, F16 = sum average, F17 = sum entropy, F18 = sum of squares variance, and F19 = sum variance, with $p$-values $< 0.00001$ for HC and $p$-values $< 0.8 \times 10^{-7}$ for PD features

| Cohort | F1 | F2 | F3 | F4 | F5 |
|---|---|---|---|---|---|
| HC | 29.6339 ± 10.5189 | 573.1053 ± 172.1820 | –12.5450 ± 30.4113 | 11.0616 ± 2.6468 | 0.1372 ± 0.1315 |
| PD | 30.5137 ± 8.8956 | 631.0476 ± 211.4640 | –20.8186 ± 24.7021 | 11.5710 ± 2.7627 | 0.1507 ± 0.1154 |
| Cohort | F6 | F7 | F8 | F9 | F10 |
| HC | 1.7429 ± 0.3568 | 5.9607 ± 1.0870 | 2.2015 ± 0.5086 | 0.1238 ± 0.1723 | 3.1969 ± 0.7053 |
| PD | 1.7467 ± 0.2990 | 6.2731 ± 1.2286 | 2.2573 ± 0.4539 | 0.1133 ± 0.1439 | 3.2030 ± 0.5959 |
| Cohort | F11 | F12 | F13 | F14 | F15 |
| HC | 0.5165 ± 0.1120 | –0.0850 ± 0.0661 | 0.4537 ± 0.1118 | 0.5746 ± 0.0984 | 0.2707 ± 0.1778 |
| PD | 0.5134 ± 0.0921 | –0.0796 ± 0.0476 | 0.4546 ± 0.0979 | 0.5717 ± 0.0810 | 0.2652 ± 0.1446 |
| Cohort | F16 | F17 | F18 | F19 | |
| HC | 10.4377 ± 2.5315 | 2.2374 ± 0.4450 | 6.3543 ± 1.3605 | 14.3587 ± 3.4359 | |
| PD | 10.6593 ± 2.1010 | 2.2621 ± 0.3803 | 6.8090 ± 1.5609 | 15.6643 ± 4.2168 | |

**Table 7.7** Means and standard deviations of 11 RQA features for healthy control (HC) and Parkinson's disease (PD) groups, with $p$-values $< 0.8 \times 10^{-6}$ for HC and $p$-values $< 0.0001$ for PD features

| Feature | HC | PD |
|---|---|---|
| $RR$ | 0.0256 ± 0.0320 | 0.0252 ± 0.0468 |
| $DET$ | 0.0523 ± 0.0614 | 0.0502 ± 0.0826 |
| $< L >$ | 2.0307 ± 0.0415 | 2.0326 ± 0.0671 |
| $L_{\max}$ | 3.2679 ± 1.3002 | 3.1833 ± 1.5784 |
| $ENT$ | 0.0166 ± 0.0080 | 0.0171 ± 0.0077 |
| $LAM$ | 0.0722 ± 0.0797 | 0.0688 ± 0.1002 |
| $TT$ | 2.0458 ± 0.0721 | 2.0455 ± 0.1053 |
| $V_{\max}$ | 3.0357 ± 1.4008 | 2.9500 ± 1.5341 |
| $T_1$ | 95.6265 ± 82.0043 | 79.3590 ± 47.8122 |
| $T_2$ | 97.3062 ± 82.2440 | 80.9654 ± 47.8072 |
| $RTE$ | 0.1265 ± 0.0222 | 0.1321 ± 0.0208 |

**Table 7.8** Mean twofold cross-validation results for classification of healthy control and early-stage Parkinson's disease subjects obtained from different methods in terms of sensitivity (SEN), specificity (SPE), area under ROC curve (AUC), and accuracy (ACC), where "N/A" indicates not available

| Method | AUC | SEN (%) | SPE (%) | ACC (%) |
|---|---|---|---|---|
| nQi1 | 0.81 | 71.00 | 84.00 | 78.00 |
| nQi2 | 0.79 | N/A | N/A | N/A |
| AFT | 0.75 | N/A | N/A | N/A |
| SKT | 0.61 | N/A | N/A | N/A |
| RQA | 0.96 | 93.50 | 90.18 | 91.90 |
| FRP-GLCM | 1 | 100 | 100 | 100 |

**Table 7.9** Variation of twofold cross-validation results for classification of healthy control and early-stage Parkinson's disease subjects obtained from RQA in terms of sensitivity (SEN), specificity (SPE), area under ROC curve (AUC), and accuracy (ACC)

| AUC | SEN (%) | SPE (%) | ACC (%) |
|---|---|---|---|
| 0.93 | 90.00 | 82.14 | 86.21 |
| 0.97 | 91.67 | 91.07 | 91.38 |
| 1 | 100 | 100 | 100 |

**Table 7.10** Cross-validation results for classification of healthy control and early-stage Parkinson's disease subjects obtained from RQA and FRP-GLCM in terms of sensitivity (SEN), specificity (SPE), area under ROC curve (AUC), and accuracy (ACC)

| Method | AUC | SEN (%) | SPE (%) | ACC (%) |
|---|---|---|---|---|
| Tenfold cross-validation | | | | |
| RQA | 0.77 | 71.67 | 71.43 | 70.00 |
| FRP-GLCM | 0.89 | 85.00 | 73.21 | 79.31 |
| LOO cross-validation | | | | |
| RQA | 0.98 | 98.33 | 96.43 | 97.41 |
| FRP-GLCM | 1 | 100 | 100 | 100 |

whereas nQi2 using each of the nonoverlapping time windows, alternating finger tapping (AFT) [69] and single key tapping (SKT) [100], are shown in Table 7.8. The 10 results of the twofold cross-validations obtained from RQA were among the three values of AUC, specificity, sensitivity, and accuracy as shown in Table 7.9. Furthermore, Table 7.10 shows the tenfold and leave-one-out (LOO) cross-validation results obtained from the RQA and FRP-GLCM.

Figure 7.20 shows time series of a healthy control and an early PD subject. Figures 7.21 and 7.22 show the FRP and three scalable recurrence networks of the corresponding time series of the HC subject and patient diagnosed with early-stage PD as shown in Fig. 7.20.

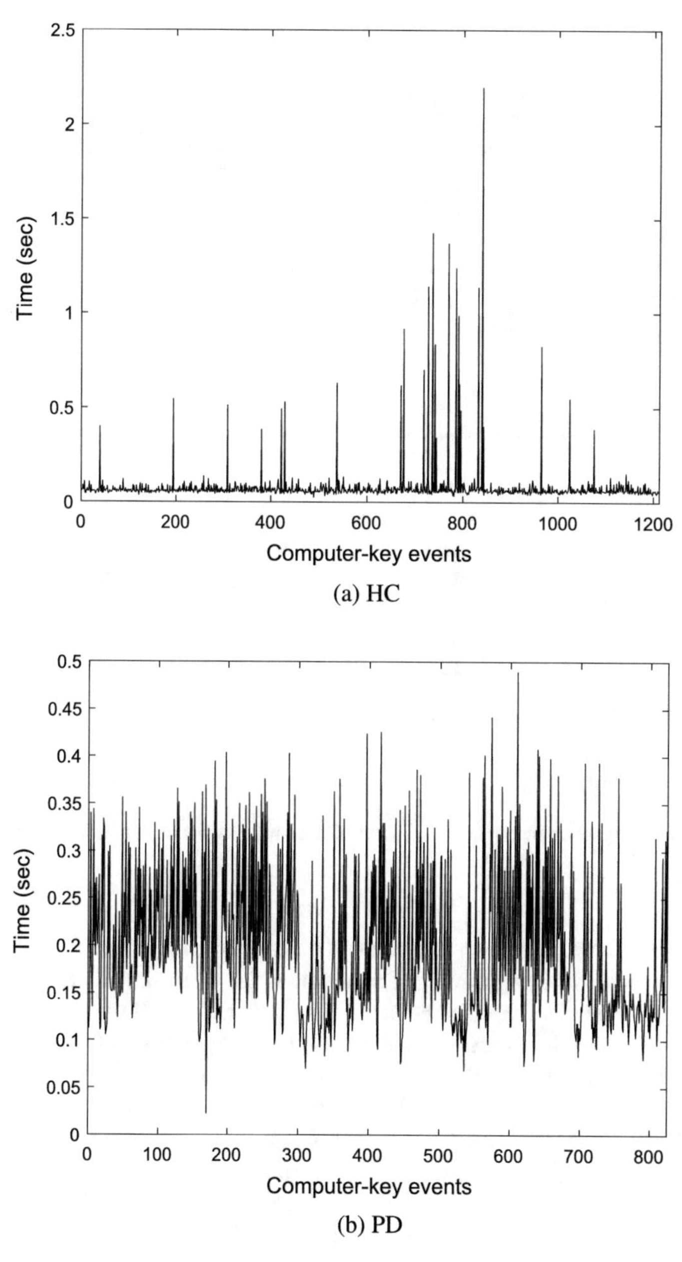


**Fig. 7.20** **a** Time series of computer-key hold time recorded from an HC and a subject with early PD

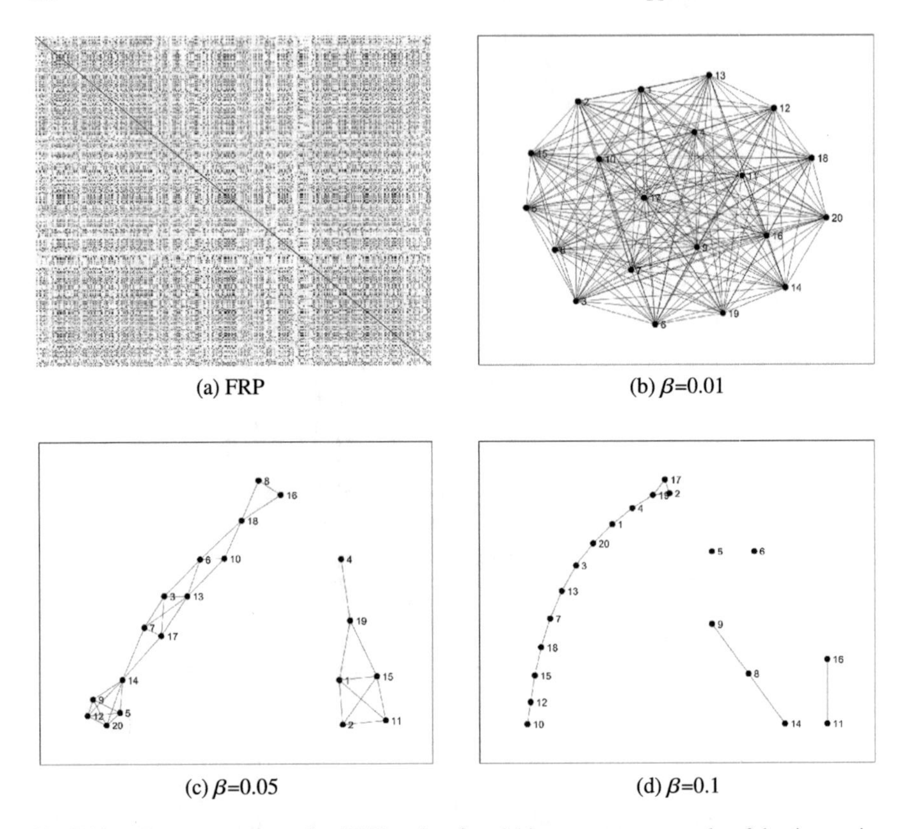

(a) FRP (b) $\beta$=0.01

(c) $\beta$=0.05 (d) $\beta$=0.1

**Fig. 7.21** **a** Fuzzy recurrence plot (FRP) and **c–d** scalable recurrence networks of the time series of an HC

Table 7.11 shows the mean values of $< CC >$ and $CPL$ obtained from scalable recurrence networks using three different values for $\beta$ for the HC and PD groups, where the $p$-values for the $< AC >$ and $CPL$ of both HC and PD groups are $< 10^{-59}$.

Based on the twofold cross-validations for the classification of 56 key-hold time series of the HC group and 60 key-hold time series of the early-stage PD group, results obtained from using either the RQA or FRP-based GLCM features for training the LS-SVM outperform those provided by the nQi model [68], alternating finger tapping (AFT) [69], and single key tapping (SKT) [100], where the AUCs = 0.81, 0.79, 0.75, 0.61, 0.96, and 1 for nQi1, nQi2, AFT, SKT, RQA, and FRP-GLCM, respectively (Table 7.8). The FRP-GLCM not only outperformed the RQA in the twofold cross-validation (Table 7.8), but also in the tenfold and LOO cross-validations (Table 7.10). Regarding the RQA features (Table 7.7), $T_1$ (recurrence time of the first type), having 96 for the HC and 79 for the PD, and $T_2$ (recurrence time of the second type), having 97 for the HC and 81 for the PD, are the most differentiating features. As for the FRP-based GLCM features (Table 7.6), F2 (cluster prominence), taking 573 for the HC and 631 for the PD, and F3 (cluster shade), taking respective negative values of

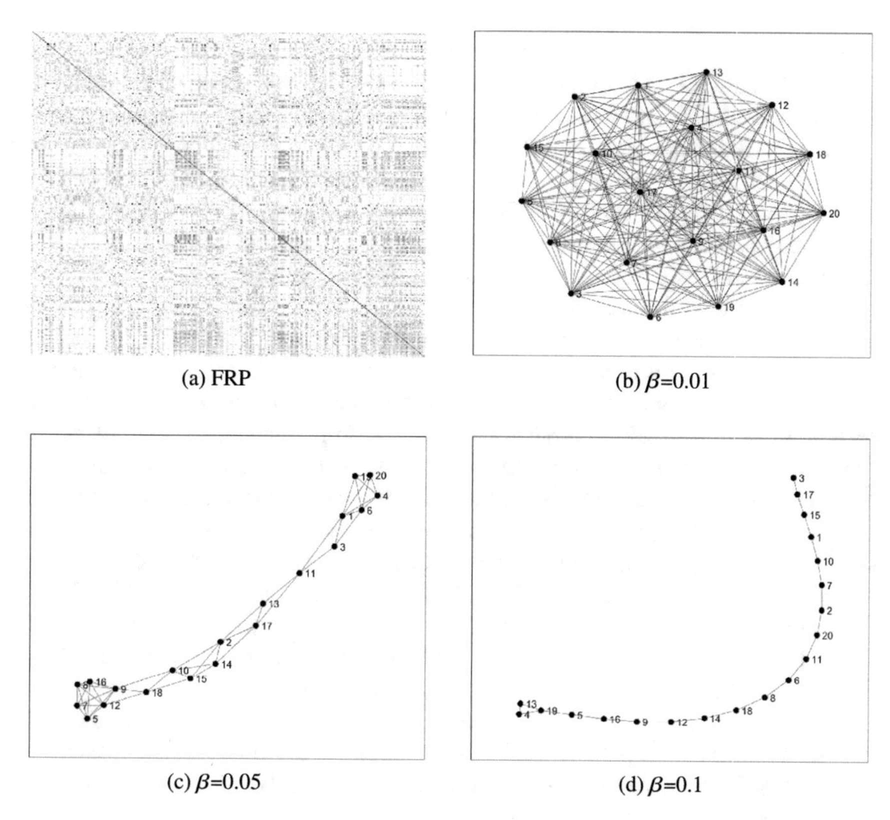


**Fig. 7.22** **a** Fuzzy recurrence plot (FRP) and **b–d** scalable recurrence networks of the time series of an early PD

**Table 7.11** Mean values of average clustering coefficient (<*CC*>) and characteristic path length (*CPL*) obtained from scalable recurrence networks using different values for $\beta$ for healthy control (HC) and Parkinson's disease (PD) groups

| *HC* | | |
|---|---|---|
| $\beta$ | <*CC*> | *CPL* |
| 0.01 | $0.9917 \pm 0.0159$ | $1.0230 \pm 0.0528$ |
| 0.05 | $0.7345 \pm 0.0335$ | $3.0935 \pm 1.8348$ |
| 0.1 | $0.1806 \pm 0.0454$ | $5.9369 \pm 3.8677$ |
| *PD* | | |
| $\beta$ | <*CC*> | *CPL* |
| 0.01 | $0.9907 \pm 0.0067$ | $1.0283 \pm 0.0509$ |
| 0.05 | $0.7094 \pm 0.0266$ | $3.3611 \pm 2.0335$ |
| 0.1 | $0.1610 \pm 0.0496$ | $6.1544 \pm 4.0275$ |

13 and 21 for the HC and PD, are the most discerning features. The $p$-values of the RQA and FRP-GLCM features for both HC and PD groups as shown in Tables 7.7 and 7.6, respectively, are statistically highly significant. The RQA features are more statistically significant than the FRP-GLCM features for the HC group, whereas the FRP-GLCM features are more statistically significant than the RQA features for the PD group. The FRP-based GLCM features are most effective for classifying the computer-keystroke time series recorded from the HC and PD groups as illustrated in the cross-validation results obtained from different methods.

Regarding visualization, the FRPs show much better visual representations of the time series than the RPs as illustrated in [70], where the similarity threshold $\epsilon$ was chosen according to a documented suggestion [101]. In fact, the selection of the number of clusters for an FRP has been found not to be as sensitive as for the selection of the similarity tolerance parameter $\epsilon$ [70], allowing the ease of use of the FRP method. For the natural selection of the embedding dimension of 1 and time delay of 1 for one-dimensional signals, the use of the recurrence networks based on recurrence plots for time series of more than 1500 sample points is not desirable for the purpose of visualization. FRPs and scalable recurrence networks not only are effective for the LS-SVM classification, but also useful for visualizing the characteristics of the time series. As a case as shown in Figs. 7.21 and 7.22, while it can be difficult to differentiate the two time series (Figs. 7.21a and 7.22a), difference in the recurrence patterns of the two groups can be more easily recognized by their FRPs (Figs. 7.21c and 7.22c), where the texture of FRP of the HC subject is finer than that of the PD patient. Similarly, topologies of the three scalable recurrence networks of the time series of the HC subject (Fig. 7.21d–f) are also different from those of the PD patient (Fig. 7.22d–f).

The values for the $< CC >$ of the HC are consistently larger than those of the PD (Table 7.11), which indicate the vertices of the scalable recurrence networks of the HC are more connected than those of the PD. The values for the $CPL$ of the HC are consistently smaller than those of the PD (Table 7.11), suggesting the vertices of the PD networks are more connected with shorter paths than those of the HC networks. The larger the value of $\beta$, the more separable the values of $< CC >$ and $CPL$ obtained for HC and PD groups. In this study, the increase of values for $\beta$ was stopped at 0.1, because a higher value would result in the generation of mostly disconnected vertices of the networks. The $p$-values for the $< CC >$ and $CPL$ of both HC and PD groups are almost zero, indicating a strong evidence of their statistical significance.

The extraction of nonlinear dynamics features using recurrence plots, fuzzy recurrence plots, and scalable recurrence networks for classifying and characterizing HC individuals from those with early-stage PD has been presented and discussed. The use of fuzzy recurrence plots with an advanced machine learning method enables the power of the classification of this type of signals. Particularly, fuzzy recurrence plots are rich in texture and therefore allow the exploration of many texture models for pattern classification, which is worth further investigation in future research applied to the analysis of complex and nonlinear keystroke time series. The number of clusters for constructing a fuzzy recurrence plot as well as values for parameter $\beta$ for constructing scalable recurrence networks from fuzzy recurrence plots was heuristically chosen in this study. The development of some analytical or numerical methods for optimal selections of these two parameters would certainly advance the applications of the proposed approach.

For the construction of either a recurrence plot or a fuzzy recurrence plot, the selection of appropriate values for the time delay and embedding dimension, which are called the embedding parameters, is naturally sought. It is reported [102] that the most commonly adopted methods for estimating the time delay and embedding dimension are the average mutual information [103] and the false nearest neighbors [104], respectively. However, these two methods are heuristic, and several other methods have been developed to provide alternative estimates for these embedding parameters [102]. It was shown that for low-dimensional systems, the same results for constructing recurrence plots can be obtained without the need for embedding [105]. Thus, the embedding parameters of one were used for the one-dimensional signals in this study. The cluster validity functions can be used for constructing scalable recurrence networks, which can be applied for selecting an appropriate number of clusters for a fuzzy recurrence plot. The number of clusters selected in this study was based on a good appearance of texture of the fuzzy recurrence plots for the purpose of texture extraction. However, being like the study of the selection of the embedding parameters, the exploration of relationship between the classification performance and the selection of the embedding parameters and number of clusters for constructing fuzzy recurrence plots from time series deserve study in its own right in future research.

Due to data availability, the proposed methods for analysis and classification of keystroke time series are limited to the study of healthy control and early-stage PD subjects. Extended applications of the proposed methods to studying PD subjects of different disease stages are certainly promising, because the pattern classification of signals measured from healthy control and early-stage PD subjects is supposed to be the most difficult task in comparison with PD subjects of more severe symptoms. In other words, patterns of motor signs in latter stages of PD would be easier for a classifier to recognize than those in an early onset of the disease. Likewise, extension of applications of the fuzzy recurrence plots and their scalable networks for differentiating physiological time series obtained from patients with PD, dementia, depression, and mixture of the diseases is promising to deliver effective results.

While the method of recurrence plots has been relatively a commonly applied method for studying time series, the potential of the newly developed method of fuzzy recurrence plots is still not widely explored for analysis of time series in physiology. Network-based analysis has been gaining increasing attention in medicine and biology [106–110]. For the first time, the method of scalable recurrence networks was adopted to investigate network properties of time series of motor signs measured from computer-key hold times of early PD. It has been suggested that new biomarker discovery together with imaging techniques, clinical examinations, and neuropathological assessments will lead to novel therapeutics for PD [111]. Comprehensive study of network-based measures addressed in this study will have an important implication as physiologic markers for early prognosis and therapeutic evaluation of the disease.

## 7.5 Visualization and Classification of Gait Dynamics

Neurodegenerative disease is a mental disorder that affects gait and mobility. Gait abnormality is a deviation from normal walking, and the observation of the pattern of a patient's walk is thought to be the most important part of the neurological assessment. Normal gait depends on the integration of strength, sensation, and coordination. It is how a person walks can reveal disorder in the nervous and musculoskeletal systems [112]. Therefore, in order to gain better understanding of the pathophysiology of these disorders and to enhance the ability for evaluating responses to therapeutic interventions, it is important to accurately quantify gait dynamics.

Evidence of different behaviors in healthy control subjects and patients with neurodegenerative disease has been found in the fluctuations in time series of stride-to-stride measures of footfall contact times. These altered stride intervals indicate changes in neurological functions associated with aging and certain disease staging, as suggested in [113] with the use of the power law to derive a scaling exponent for elderly, young, and Huntington's disease (HD), which is disorder that causes the progressive degeneration of nerve cells in the brain. The scaling exponent was found to be higher in young subjects than elderly subjects, and higher in healthy control (HC) group than the patients with HD. Altered gait rhythm was found in patients with amyotrophic lateral sclerosis (ALS), which is a neural disorder caused by the loss of motoneurons, and evidence of altered stride dynamics was also found in advanced ALS, HD, and Parkinson's disease (PD) [114], which is a progressive degeneration of the nervous system that affects the motor system or movement.

Based on the findings reported in [113, 114] and the available associated gait datasets, several studies have been attempted to extract new features of the time series of the gait rhythm for pattern analysis and classification of HC, PD, HD, and ALS subjects. Wu and Krishnan [115] applied the Parzen window method to estimate the probability density function of the stride intervals and its subphases, which are the swing and stance intervals, and the signal turn counts of the time series to characterize and classify HC and PD groups using the least squares support vector machines. In a similar study, Khorasani and Daliri [116] applied the hidden Markov models for classifying the time series of the right stride interval of HC and PD subjects. Zeng and Wang [117] applied the theory of dynamic learning to classify HC, PD, HD, and ALS subjects using features extracted from the left and right swing intervals and the left and right stance intervals, where the work in [64] used a different database for the classification of HC and PD subjects based on the theory of dynamic learning. Most recently, Ren et al. [65] studied the gait fluctuations of HC, PD, HD, and ALS using the empirical mode decomposition to extract two parameters for characterizing the four groups, and used several classifiers trained with these parameters to carry out the classification of the four groups.

This work [66] attempted to extract texture features from time series of gait dynamics in HC, PD, HD, and ALS by transforming the one-dimensional signals (time series) into two-dimensional texture objects (images) using the method of fuzzy recurrence plots (FRPs). The FRPs not only can be useful as a visualization tool but their textural information can be readily extracted by texture analysis methods for pattern analysis and classification. Here, the gray-level co-occurrence matrix (GLCM) method is utilized to obtain various texture features of the FRPs because of the similarity in the underlying formulations of FRP and GLCM methods (both methods work on capturing information about recurrences of variables in space). State-of-the-art machine learning method trained with the GLCM-based texture features was applied for differentiating HC subjects from the three groups of patients with neurodegenerative diseases.

### 7.5.1 Database

The Gait Dynamics in Neurodegenerative Disease Database [113, 114] was used in this study. The database consists of the raw data obtained by using force-sensitive resistors, where the output being proportional to the force under the foot, of healthy control (HC) subjects and three types of neurodegenerative disease: Parkinson's disease (PD), Huntington's disease (HD), and amyotrophic lateral sclerosis (ALS). The numbers of control subjects = 16, PD patients = 15, HD patients = 20, and ALS patients = 13. Stride-to-stride measures of footfall contact times were derived from these signals.

Tables 7.12, 7.13, 7.14, and 7.15 show the clinical information for subjects in the HC, PD, HD, and ALS groups, including age, gender, height, weight, walking speed, and a measure of disease severity (PD and HD) or duration (ALS). For the PD patients, the Hoehn and Yahr scale of stages 1 through 5 [118] is used, where a higher scale indicates more advanced disease and "x" indicates a missing value. For the HD patients, the total functional capacity measure is applied, where a lower score indicates more advanced functional impairment. For the ALS patients, the value is the time in months since the diagnosis of the disease. For the control subjects, an indicator of 0 is used.

### 7.5.2 Results

Only the right-foot stride-interval signals of the HC, PD, HD, and ALS cohorts were used in this study for constructing FRPs and extracting texture features for pattern classification. The parameters for constructing the FRPs were selected as follows: embedding dimension =1, time delay = 1, the number of clusters = 5, where the maximum number of iterations and the minimum value for improvement in the objective function between two consecutive iterations of the FCM algorithm were 100 and 0.00001, respectively, and the fuzzy weighting exponent = 2.

The stride intervals are the durations measured in seconds from the time the foot contacted the ground to the next ground contact of the same foot. The raw signals

**Table 7.12** Clinical information for subjects in the healthy control (HC) group

| Subject # | Age (years) | Height (m) | Weight (kg) | Gender | Gait speed (m/s) | Severity |
|---|---|---|---|---|---|---|
| 1 | 57 | 1.94 | 95 | F | 1.33 | 0 |
| 2 | 22 | 1.94 | 70 | M | 1.47 | 0 |
| 3 | 23 | 1.83 | 66 | F | 1.44 | 0 |
| 4 | 52 | 1.78 | 73 | F | 1.54 | 0 |
| 5 | 47 | 1.94 | 82 | F | 1.54 | 0 |
| 6 | 30 | 1.81 | 59 | F | 1.26 | 0 |
| 7 | 22 | 1.86 | 64 | F | 1.54 | 0 |
| 8 | 22 | 1.78 | 64 | F | 1.33 | 0 |
| 9 | 32 | 1.83 | 68 | F | 1.47 | 0 |
| 10 | 38 | 1.67 | 57 | F | 1.4 | 0 |
| 11 | 69 | 1.72 | 68 | F | 0.91 | 0 |
| 12 | 74 | 1.89 | 77 | M | 1.26 | 0 |
| 13 | 61 | 1.86 | 60 | F | 1.33 | 0 |
| 14 | 20 | 1.9 | 57 | F | 1.33 | 0 |
| 15 | 20 | 1.83 | 50 | F | 1.19 | 0 |
| 16 | 40 | 1.74 | 59 | F | 1.33 | 0 |

**Table 7.13** Clinical information for subjects in the Parkinson's disease (PD) group

| Subject # | Age (years) | Height (m) | Weight (kg) | Gender | Gait speed (m/s) | Severity |
|---|---|---|---|---|---|---|
| 1 | 77 | 2.00 | 86 | M | 0.98 | 4 |
| 2 | 44 | 1.67 | 54 | F | 1.26 | 1.5 |
| 3 | 80 | 1.81 | 77 | M | 0.98 | 2 |
| 4 | 74 | 1.72 | 43 | F | 0.91 | 3.5 |
| 5 | 75 | 1.92 | 91 | M | 1.05 | 2 |
| 6 | 53 | 2.00 | 86 | M | 1.33 | 2 |
| 7 | 64 | 1.67 | 54 | F | 0.91 | 4 |
| 8 | 64 | 1.83 | 73 | M | 0.84 | 4 |
| 9 | 68 | 1.92 | 84 | M | 1.05 | 1.5 |
| 10 | 60 | 1.94 | 74 | M | 1.19 | 3 |
| 11 | 74 | 2.04 | 100 | M | 0.5 | 3 |
| 12 | 57 | 1.72 | 65 | F | 0.98 | 3 |
| 13 | 79 | 1.68 | 59 | F | 0.84 | 3 |
| 14 | 57 | 2.13 | 84 | M | 0.98 | 3 |
| 15 | 76 | 2.00 | 96 | M | 1.19 | 2.5 |

**Table 7.14** Clinical information for subjects in the Huntington's disease (HD) group

| Subject # | Age (years) | Height (m) | Weight (kg) | Gender | Gait speed (m/s) | Severity |
|---|---|---|---|---|---|---|
| 1 | 42 | 1.86 | 72 | M | 1.68 | 8 |
| 2 | 41 | 1.78 | 58 | F | 1.05 | 11 |
| 3 | 66 | 1.75 | 63 | F | 1.05 | 4 |
| 4 | 47 | 1.88 | 64 | F | 1.4 | 2 |
| 5 | 36 | 2.00 | 85 | M | 1.82 | 10 |
| 6 | 41 | 1.83 | 59 | F | 1.54 | 8 |
| 7 | 71 | 2.00 | 75 | M | 1.05 | 2 |
| 8 | 53 | 1.81 | 56 | F | 1.26 | 9 |
| 9 | 54 | 1.8 | 90 | F | 1.26 | 12 |
| 10 | 47 | 1.78 | 102 | F | 1.05 | 4 |
| 11 | 33 | 1.97 | 84 | M | 1.26 | 11 |
| 12 | 47 | 1.92 | 75 | M | 1.19 | 8 |
| 13 | 40 | 1.72 | 48 | F | 0.56 | 5 |
| 14 | 36 | 1.88 | 97 | F | 1.4 | 12 |
| 15 | 34 | 1.94 | 88 | F | 0.56 | 3 |
| 16 | 70 | 1.83 | 93 | M | 0.56 | 5 |
| 17 | 29 | 1.78 | 76 | F | 1.19 | 12 |
| 18 | 54 | 1.72 | 53 | F | 0.98 | 2 |
| 19 | 59 | 1.78 | 58 | F | 0.98 | 1 |
| 20 | 33 | 1.57 | 45 | F | x | 9 |

**Table 7.15** Clinical information for subjects in the amyotrophic lateral sclerosis (ALS) group

| Subject # | Age (years) | Height (m) | Weight (kg) | Gender | Gait speed (m/s) | Duration (months) |
|---|---|---|---|---|---|---|
| 1 | 68 | 1.80 | 86.18 | M | 1.302 | 1 |
| 2 | 63 | 1.83 | 83.92 | M | 1.219 | 14 |
| 3 | 70 | 1.57 | 40.82 | F | 0.853 | 13 |
| 4 | 70 | 1.70 | 58.97 | F | x | 54 |
| 5 | 36 | 1.70 | 74.39 | M | x | 5.5 |
| 6 | 43 | 1.75 | 68.95 | M | 0.77 | 17 |
| 7 | 65 | 1.73 | 81.65 | M | 1.302 | 9 |
| 8 | 51 | 1.83 | 106.6 | M | 1.085 | 3 |
| 9 | 50 | 1.58 | 61.24 | M | 0.899 | 54 |
| 10 | 40 | 1.70 | 61.24 | F | 1.219 | 14.5 |
| 11 | 39 | 1.88 | 83.92 | M | 1.283 | 7 |
| 12 | 62 | 1.78 | 117.5 | M | 0.831 | 12 |
| 13 | 66 | 1.83 | x | M | 0.832 | 34 |

were smoothed using the third-order one-dimensional median filter. Figures 7.23 and 7.24 show examples of the right-foot stride-interval signals and corresponding FRPs of an HC, a PD, an HD, and an ALS subject. The RP of the HC has the strongest indication of recurrences among the other three subjects. The RP of the ALS has the most contrast of recurrent (very dark bands) and nonrecurrent (very light bands) patterns. The image intensities of the RPs of the PD and HD subjects are similar. But the RP of the HD consists of many small dark and light bands, while the RP of the PD has more grayish or blurry sub-structures. It was reported that the FRPs visually exhibit similar textures of the pairs of gait signals in the same groups, and difference in texture patterns of the FRPs of the pairs of gait signals between the four groups can be observed [66].

There are 19 GLCM-based features extracted from the FRPs. The orientation for computing the GLCM is one pixel to the right, which indicates $d = 1$ and $\theta = [0, 1]$. The means and standard deviations of the 19 GLCM-based features are shown in Table 7.16, where T1 = autocorrelation, T2 = cluster prominence, T3 = cluster shade, T4 = contrast, T5 = correlation, T6 = difference entropy, T7 = difference variance, T8 = dissimilarity, T9 = energy, T10 = entropy, T11 = homogeneity, T12 = information measure of correlation 1, T13 = information measure of correlation 2, T14 = inverse difference, T15 = maximum probability, T16 = sum average, T17 = sum entropy, T18 = sum of squares variance, and T19 = sum variance.

The six texture features that well distinguish HC from PD, HD, and ALS groups are T1, T2, T3, T11, T17, and T18. Particularly, the cluster shade (T3) is a measure of the skewness of the GLCM and its purpose is to quantify the perceptual concepts of uniformity [119], and cluster prominence (T2) quantifies the variation in grayscale as found in sonographic features between the normal and postradiotherapy parotid glands [120]. The mean values of T14 (inverse difference) of the four groups are the same ($0.0008 \times 10^3$) but having different standard deviations. The standard deviations of the textures of the HC group are smallest, and the standard deviations of the textures of the PD group are smaller than those of both HD and ALS groups. This is in agreement with the finding that in healthy control subjects, the variance of gait dynamics is relatively small in comparison with PD, HD, and ALS subjects [114]. The $p$-values $< 0.000001$ were obtained for all texture features, except the $p$-value of T3 (cluster shade) of the ALS group = 0.0064, which demonstrate the statistical significance of the texture features of the four groups.

To differentiate HC subjects from PD, HD, and ALS subjects, the Least Squares Support Vector Machines (LS-SVM) toolbox (LS-SVMlab v1.8) [99] was used for the classification task. Using the leave-one-out (LOO) cross-validation, the LS-SVM trained and tested with the GLCM-based features extracted from the FRPs, denoted as FRP (LS-SVM), of the gait signals achieved 100% for the accuracy (ACC), sensitivity (SEN), and specificity (SPE), and 1 (maximum value) for the area under curve (AUC) of the receiver operating characteristic (ROC) curve in all three binary classification cases of HC and PD, HC and HD, and HC and ALS.

Using the same database, the work reported in [115] used both stance and swing intervals to obtain the standard deviations of the probability density functions with the Parzen window method, and the stride interval, swing interval, and stance interval to extract the signal turn count (STC) parameters to classify HC and PD with the

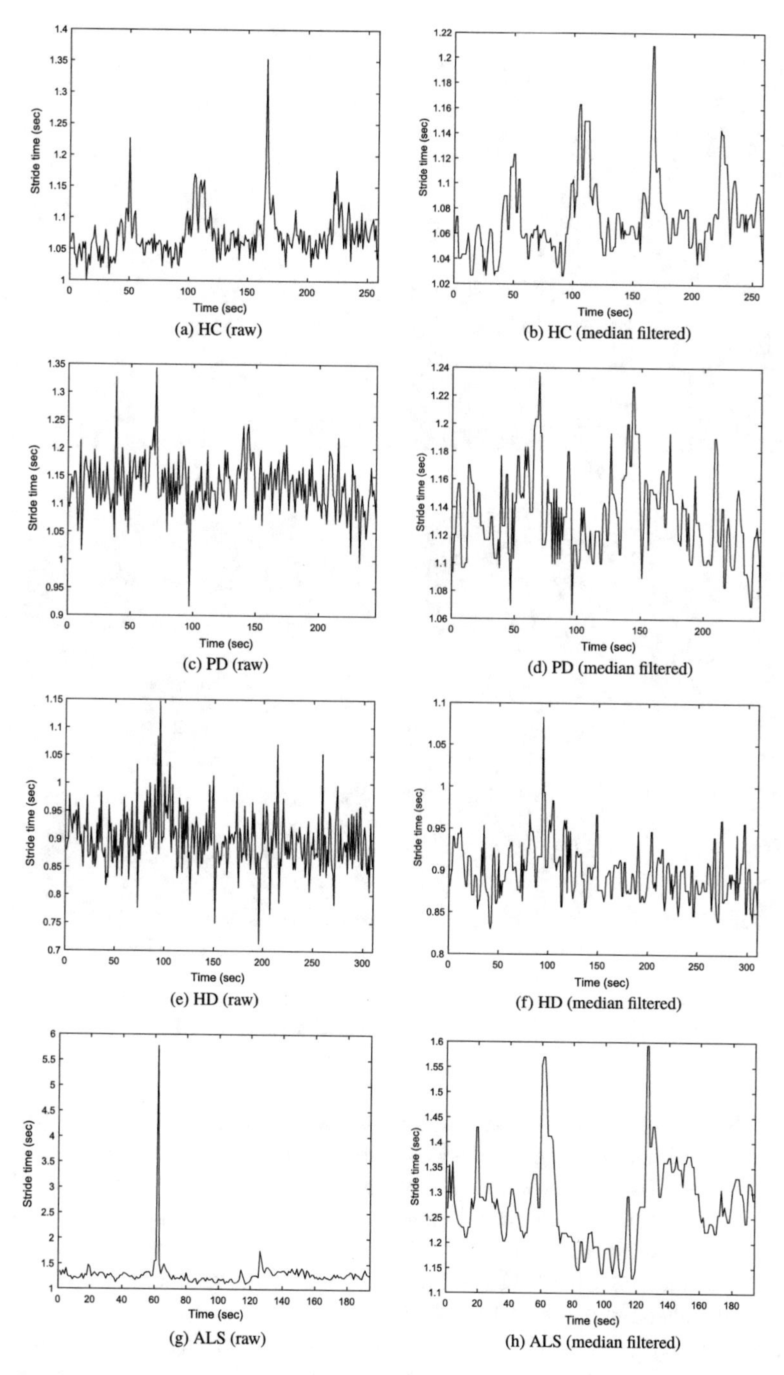


**Fig. 7.23** Gait dynamics of healthy control (HC), Parkinson's disease (PD), Huntington's disease (HD), and amyotrophic lateral sclerosis (ALS)

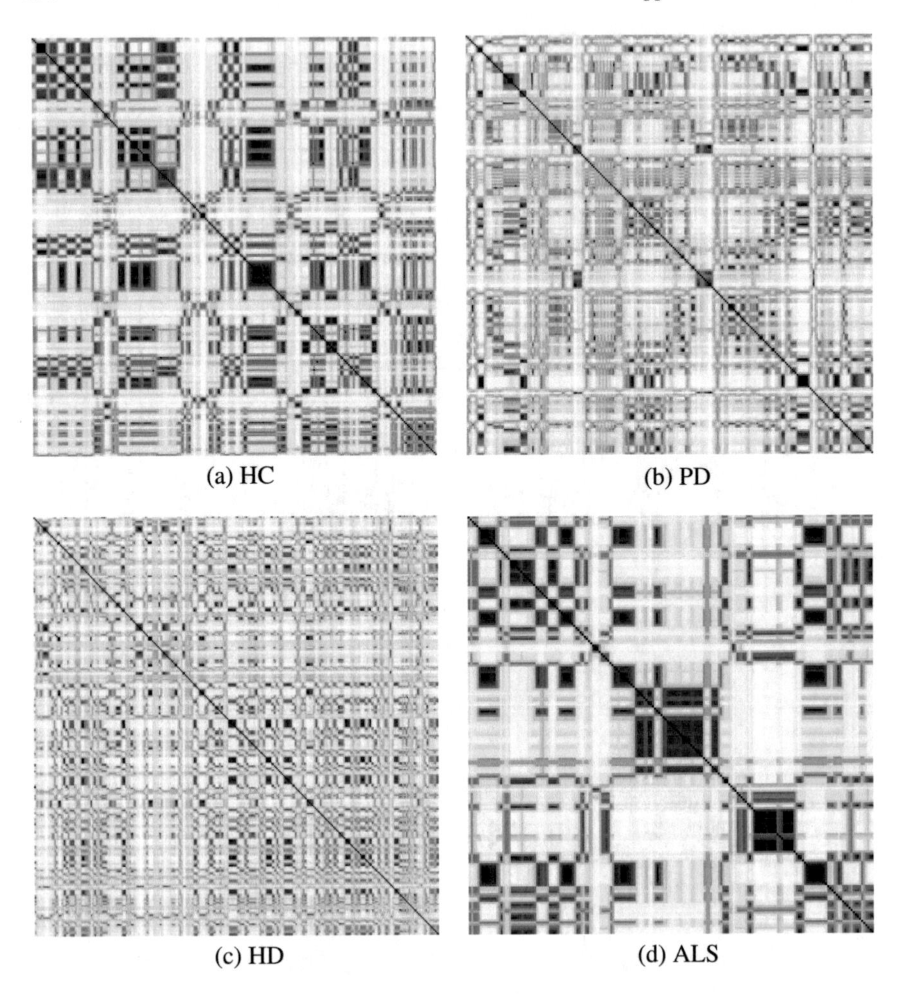


**Fig. 7.24** Fuzzy recurrence plots of gait dynamics of healthy control (HC), Parkinson's disease (PD), Huntington's disease (HD), and amyotrophic lateral sclerosis (ALS)

same LS-SVM toolbox. The AUC and accuracy of this method obtained from the LOO cross-validation using the LS-SVM classifier, denoted as $\sigma$ & STC (LS-SVM), are 0.952 and 90.32%, respectively. Another report [116] using continuous hidden Markov models of the raw data, denoted as RD (HMM), of the right-foot stride intervals of the same HC and PD subjects to carry out the classification task achieved the same accuracy rate of 90.32% as the $\sigma$ & STC method, with sensitivity = 93.33% (14/15) and specificity = 87.50% (14/16).

The work reported in [117] applied a deterministic learning method, denoted as DL, to extract four features based on the information of the average swing, stance, and stride intervals, and the fluctuations of the time series, all with both left and right feet. The LOO cross-validation of these features using the DL method obtained

**Table 7.16** Means ($\mu$) and standard deviations ($\sigma$) of GLCM-based texture features obtained from FRPs of healthy control (HC), Parkinson's disease (PD), Huntington's disease (HD), and amyotrophic lateral sclerosis (ALS) groups

| Texture | HC | | PD | | HD | | ALS | |
|---|---|---|---|---|---|---|---|---|
| | $\mu$ ($\times 10^3$) | $\sigma$ | $\mu$ ($\times 10^3$) | $\sigma$ | $\mu$ ($\times 10^3$) | $\sigma$ | $\mu$ ($\times 10^3$) | $\sigma$ |
| T1 | 0.0410 | 1.8091 | 0.0414 | 3.3251 | 0.0428 | 2.2179 | 0.0390 | 7.9936 |
| T2 | 1.0544 | 83.1435 | 1.0758 | 83.3166 | 1.1147 | 495.1626 | 1.2209 | 281.8654 |
| T3 | –0.0831 | 9.7143 | –0.0836 | 13.4509 | –0.0920 | 40.4738 | –0.0617 | 67.4312 |
| T4 | 0.0052 | 0.8822 | 0.0048 | 1.1958 | 0.0052 | 1.5167 | 0.0043 | 1.3465 |
| T5 | 0.0006 | 0.0618 | 0.0006 | 0.1039 | 0.0006 | 0.1308 | 0.0007 | 0.1204 |
| T6 | 0.0012 | 0.0865 | 0.0012 | 0.1025 | 0.0012 | 0.2754 | 0.0011 | 0.1864 |
| T7 | 0.0040 | 0.5978 | 0.0037 | 0.7970 | 0.0040 | 1.0633 | 0.0034 | 0.8417 |
| T8 | 0.0011 | 0.1452 | 0.0010 | 0.1953 | 0.0011 | 0.2900 | 0.0009 | 0.2596 |
| T9 | 0.0002 | 0.0432 | 0.0002 | 0.0395 | 0.0002 | 0.0998 | 0.0002 | 0.0633 |
| T10 | 0.0026 | 0.1792 | 0.0026 | 0.1648 | 0.0025 | 0.4604 | 0.0024 | 0.2652 |
| T11 | 0.0007 | 0.0278 | 0.0008 | 0.0329 | 0.0008 | 0.0626 | 0.0008 | 0.0550 |
| T12 | –0.0003 | 0.0401 | –0.0003 | 0.0636 | –0.0003 | 0.1502 | –0.0004 | 0.1003 |
| T13 | 0.0008 | 0.0361 | 0.0008 | 0.0569 | 0.0008 | 0.0503 | 0.0008 | 0.0669 |
| T14 | 0.0008 | 0.0239 | 0.0008 | 0.0286 | 0.0008 | 0.0551 | 0.0008 | 0.0479 |
| T15 | 0.0004 | 0.0516 | 0.0004 | 0.0485 | 0.0005 | 0.0789 | 0.0005 | 0.0888 |
| T16 | 0.0122 | 0.3220 | 0.0122 | 0.6735 | 0.0125 | 0.3087 | 0.0116 | 1.8441 |
| T17 | 0.0021 | 0.1261 | 0.0020 | 0.1172 | 0.0019 | 0.3338 | 0.0020 | 0.1751 |
| T18 | 0.0064 | 0.3393 | 0.0063 | 0.6234 | 0.0062 | 0.6586 | 0.0068 | 1.4324 |
| T19 | 0.0203 | 1.1838 | 0.0203 | 2.8417 | 0.0195 | 3.7464 | 0.0230 | 6.1030 |

for classifying HC and PD is 87.10%, with sensitivity = 86.67% and specificity = 86.50%. Most recently, the empirical mode decomposition (EMD) [65] was used to extract features of five gait signals (stride interval, swing interval, stance interval, percentage swing interval, and percentage stance interval) to obtain the Kendall coefficients of concordance and ratios for energy change, which were used for classification by the random forest, denoted as EMD (RF); simple logistic regression, denoted as EMD (SLR); multilayer perceptron, denoted as EMD (MLP); Naive Bayes, denoted as EMD (NB); and support vector machine, denoted as EMD (SVM). The AUCs obtained from the five classifiers EMD (RF), EMD (SLR), EMD (MLP), EMD (NB), and EMD (SVM) are 0.865, 0.949, 0.910, 0.875, and 0.906, respectively. Those results are shown in Table 7.17. The $\sigma$ & STC features [115] were also used to differentiate HC from PD subjects using the linear discriminant analysis (LDA), where the LOO cross-validation = 67.74%. The LOO cross-validation of the FRP using LDA is 77.42%. Table 7.18 shows these comparative results for the classification of the HC and PD subjects.

**Table 7.17** Comparisons of receiver operating characteristics and LOO cross-validation results for classification of HC and PD subjects, where “x” indicates not available

| Method | SEN (%) | SPE (%) | AUC | ACC (%) |
|---|---|---|---|---|
| $\sigma$ & STC (LS-SVM) [115] | x | x | 0.952 | 90.32 |
| RD (HMM) [116] | 93.33 | 87.50 | x | 90.32 |
| EMD (RF) [65] | x | x | 0.865 | x |
| EMD (SLR) [65] | x | x | 0.949 | x |
| EMD (MLP) [65] | x | x | 0.910 | x |
| EMD (NB) [65] | x | x | 0.875 | x |
| EMD (SVM) [65] | x | x | 0.906 | x |
| DL [117] | 86.67 | 86.50 | x | 87.10 |
| FRP (LS-SVM) | 100 | 100 | 1 | 100 |

**Table 7.18** Comparisons of LOO cross-validation results from LDA-based classification of HC and PD subjects

| Method | ACC (%) |
|---|---|
| $\sigma$ & STC [115] | 67.74 |
| FRP | 77.42 |

For the LOO cross-validation of the HC and HD groups, the DL method [117] provided the accuracy = 83.33%, sensitivity = 85.00%, and specificity = 81.25%. The AUCs obtained from the five classifiers EMD (RF), EMD (SLR), EMD (MLP), EMD (NB), and EMD (SVM) [65] are 0.885, 0.843, 0.878, 0.898, and 0.900, respectively. Table 7.19 shows the comparative classification results obtained from the proposed FRP and other methods (DL and EMD).

For the LOO cross-validation of the HC and ALS groups, the DL method [117] provided the accuracy = 89.66%, sensitivity = 92.31%, and specificity = 87.50%. The AUCs obtained from the five classifiers EMD (RF), EMD (SLR), EMD (MLP), EMD (NB), and EMD (SVM) [65] are 0.900, 0.859, 0.934, 0.891, and 0.906, respectively. Table 7.20 shows the comparative classification results obtained from the proposed FRP and other methods (DL and EMD).

In summary, Tables 7.17, 7.18, 7.19, and 7.20 indicate that the proposed method provides the best results over several other methods for classifying patterns of gait dynamics between HC subjects and PD, HD, and ALS patients.

**Table 7.19** Comparisons of receiver operating characteristics and LOO cross-validation for classification of HC and HD subjects, where "x" indicates not available

| Method | SEN (%) | SPE (%) | AUC | ACC (%) |
|---|---|---|---|---|
| EMD (RF) [65] | x | x | 0.885 | z |
| EMD (SLR) [65] | x | x | 0.843 | x |
| EMD (MLP) [65] | x | x | 0.878 | x |
| EMD (NB) [65] | x | x | 0.898 | x |
| EMD (SVM) [65] | x | x | 0.900 | x |
| DL [117] | 85.00 | 81.25 | x | 83.33 |
| FRP (LS-SVM) | 100 | 100 | 1 | 100 |

**Table 7.20** Comparisons of receiver operating characteristics and LOO cross-validation for classification of HC and ALS subjects, where "x" indicates not available

| Method | SEN (%) | SPE (%) | AUC | ACC (%) |
|---|---|---|---|---|
| EMD (RF) [65] | x | x | 0.900 | x |
| EMD (SLR) [65] | x | x | 0.859 | x |
| EMD (MLP) [65] | x | x | 0.934 | x |
| EMD (NB) [65] | x | x | 0.891 | x |
| EMD (SVM) [65] | x | x | 0.906 | x |
| DL [117] | 92.31 | 87.50 | x | 89.66 |
| FRP (LS-SVM) | 100 | 100 | 1 | 100 |

A procedure for transforming physiological time series into images of rich texture using the FRP method has been presented and applied for characterization and classification of gait dynamics of healthy control and the three neurodegenerative diseases. The FRPs can be useful for visualizing textural patterns of physiological signals. The extraction of texture features from the FPRs using the GLCM fits well into the underlying mechanism for capturing information content from data carried out by the two methods. The combination of the GLCM-based texture features extracted from the FRPs of the time series of the right-foot stride intervals and the LS-SVM classifier is promising for differentiating healthy control subjects from patients with neurodegenerative disease.

## 7.6 Recurrence of White Matter Lesions on MRI

White matter lesions (WMLs) shown on magnetic resonance imaging (MRI) are regions of demyelinated cells found in the white matter of the brain. Minor cases of WMLs are commonly found in people over 65 years old as the result of normal aging [121]. However, WMLs are hypothesized to be related to various geriatric disorders including progressive neurological diseases that cause brain degeneration such as multiple sclerosis and Alzheimer's disease, major depressive disorder, cardiovascular diseases, and psychiatric disorders [121]. While it is not fully understood about the mechanism of WMLs in causing brain dysfunction, WML can be considered as a biomarker for underlying pathology. Furthermore, the accurate quantification of WMLs is an important frontal result used for validating the hypothesis of the associations of WMLs and diseases. Current measures of WMLs involve qualitative or semiquantitative visual rating scales. However, these scales of WMLs are likely subject to inconsistency [122] and limited due to the nonlinearity of data as well as the lack of sensitivity to small changes in the hyperintensities of WMLs [123].

A 2D nonlinear dynamics method for an objective quantification of the WMLs is presented in this section [124]. The method works by constructing a 2D fuzzy weighted recurrence network (2D-FWRN) for quantifying WMLs on MRI scans. Network properties can be calculated for the constructed networks. These network properties are then used to express different patterns of the WMLs as values of continuous quantity.

### 7.6.1 2D-FWRN

Based on the formulation of the MC-FWRN described in Chap. 5, a 2D-FWRN is described as follows [124].

Let $\mathbf{I} = [f_{ij}]$ be a 2D image of size $M \times N$, where $i = 1, \ldots, M$, $j = 1, \ldots, N$. Let $m \geq 1$ be an integer, a local image window $\mathbf{W}_{ij}^{k} \in \mathbf{I}$ of size $(2m+1) \times (2m+1)$ is constructed for each pixel located at $ij$ of the image, where $ij$ is the center of the window. This window can be considered as embedding dimensions in two-dimensional space, which considers the local spatial distribution around $f_{ij}$ of the image. The Frobenius norm can be used to transform each local window into a scalar measure that has the useful property of invariance under rotations as

$$\|\mathbf{W}_{ij}\|_F = \sqrt{\sum_{i-m}^{i+m} \sum_{j-m}^{j+m} |f_{ij}|^2}, \tag{7.10}$$

where $(i-m), (j-m) > 0$, $(i+m) \leq M$, $(j+m) \leq N$.

We can then obtain a set of $\|\mathbf{W}_{ij}\|_F$ over the image as a set of feature scalars. Since the Frobenius norm-induced feature set can be computed for image $\mathbf{I}$, the 2D-FWRN of the image can be constructed as follows. To simplify the notation in subsequent mathematical presentation, $\|\mathbf{W}_{ij}\|_F$ is now denoted as $\mathbf{x}_n$, $n = 1, \ldots, L$, where $L$ is the total number of the feature scalars, and some same indices are used but defined differently.

Let $\mathbf{X} = \{\mathbf{x}_n\}$, $n = 1, \ldots, L$, $c$ a given number of clusters of the feature space, and a set of $c$ fuzzy clusters, $\mathbf{V} = \{\mathbf{v}_i : i = 1, \ldots, c\}$. Fuzzy clusters are groups that contain data points, where every data point has a degree of fuzzy membership belonging to each group. A fuzzy relation $\tilde{\mathbf{R}}$ between $\mathbf{v}_i$ and $\mathbf{v}_j$, $i, j = 1, \ldots, c$, is characterized by a fuzzy membership function $\mu \in [0, 1]$, which expresses the degree of similarity of each pair $(\mathbf{v}_i, \mathbf{v}_j)$ in $\tilde{\mathbf{R}}$. This fuzzy relation has the following three properties:

1. Reflexivity: $\mu(\mathbf{v}_i, \mathbf{v}_i) = 1, \ \forall \mathbf{v}_i \in \mathbf{V}$.
2. Symmetry: $\mu(\mathbf{v}_i, \mathbf{x}_n) = \mu(\mathbf{x}_n, \mathbf{v}_i), \ \forall \mathbf{x}_n \in \mathbf{X}, \forall \mathbf{v}_i \in \mathbf{V}$.
3. Transitivity: $\mu(\mathbf{v}_i, \mathbf{v}_j) = \vee_{\mathbf{x}_n}[\mu(\mathbf{v}_i, \mathbf{x}_n) \wedge \mu(\mathbf{v}_j, \mathbf{x}_n)], \ \forall \mathbf{v}_i, \mathbf{v}_j \in \mathbf{V}$, where the symbols $\vee$ and $\wedge$ stand for max and min, respectively.

An $N \times N$ adjacency matrix of a 2D-FWRN can be constructed with the associated fuzzy relation as

$$\mathbf{W} = \tilde{\mathbf{R}} - \mathbf{I}, \tag{7.11}$$

where $\mathbf{W}$ is an $N \times N$ adjacency matrix of edge weights, $\mathbf{I}$ is the $N \times N$ identity matrix, and $\tilde{\mathbf{R}}$ is also known as an FRP.

### 7.6.2 Results

Tables 7.21 and 7.22 show the values of the CC and CP for the mild, medium, and severe ratings of WMLs on MRI with $m = 1$ and $m = 3$ for various values of $c$, respectively. Based on the results shown on the two tables, the values of the two network properties consistently increase with the severity of the rating. In other words, the network properties consistently show the highest values for the severe rating and the lowest for the mild rating. The FRPs of the severe and mild WMLs are more similar to each other than the FRP of the mild WML, which are also reflected with the differences in values of the two network properties between the three WML patterns.

The MRI scans (T2 weighted FLAIR) of three participants taken from the data reported in [125] were used in this study. The WMLs on these three scans as shown in Fig. 7.25 are clinically rated as mild, medium, and severe, respectively [125]. The 2D-FWRN method was carried out to determine the clustering coefficient (CC) and characteristic path length (CP) of each of the scans. Using the partition coefficient as a measure of cluster validity described in Chap. 3, an indication of $c = 20$ was obtained. To test the consistency of the network properties using the 2D-FWRN method, $m = 1$ and 3, and $c$ as 10, 15, 20, 25, and 30. The FRPs of the MRI scans, which were cropped out to maximize the elimination of the non-MML background, were constructed with $m = 1$ and $c = 20$.

It can be observed from Tables 7.21 and 7.22 that the larger the number of clusters are used to construct the weighted network, the smaller the values of the two network properties are obtained. The CP tends to better differentiate between the WML patterns than the CC. The smaller the number of clusters is, the bigger the values of the two network properties are.

**Table 7.21** Values of average clustering coefficient (CC) and characteristic path length (CP) of 2D-FWRNs of MRI scans, where $m = 1$ and different numbers of clusters ($c$)

| | Mild | Medium | Severe |
|---|---|---|---|
| $c = 10$ | | | |
| CC | 0.0348 | 0.0362 | 0.0384 |
| CP | 0.0324 | 0.0361 | 0.0428 |
| $c = 15$ | | | |
| CC | 0.0202 | 0.0223 | 0.0244 |
| CP | 0.0177 | 0.0217 | 0.0262 |
| $c = 20$ | | | |
| CC | 0.0150 | 0.0169 | 0.0183 |
| CP | 0.0141 | 0.0175 | 0.0201 |
| $c = 25$ | | | |
| CC | 0.0121 | 0.0136 | 0.0150 |
| CP | 0.0117 | 0.0146 | 0.0171 |
| $c = 30$ | | | |
| CC | 0.0104 | 0.0123 | 0.0127 |
| CP | 0.0104 | 0.0145 | 0.0151 |

**Table 7.22** Values of average clustering coefficients (CC) and characteristic path lengths (CP) of 2D-FWRNs of MRI scans, where $m = 3$ and different numbers of clusters ($c$)

| | Mild | Medium | Severe |
|---|---|---|---|
| $c = 10$ | | | |
| CC | 0.0349 | 0.0371 | 0.0380 |
| CP | 0.0329 | 0.0373 | 0.0402 |
| $c = 15$ | | | |
| CC | 0.0200 | 0.0225 | 0.0234 |
| CP | 0.0170 | 0.0227 | 0.0244 |
| $c = 20$ | | | |
| CC | 0.0147 | 0.0169 | 0.0178 |
| CP | 0.0131 | 0.0176 | 0.0196 |
| $c = 25$ | | | |
| CC | 0.0118 | 0.0136 | 0.0144 |
| CP | 0.0110 | 0.0144 | 0.0163 |
| $c = 30$ | | | |
| CC | 0.0103 | 0.0118 | 0.0125 |
| CP | 0.0102 | 0.0131 | 0.0150 |

**Fig. 7.25** Brain MRI scans showing WMLs and associated ratings

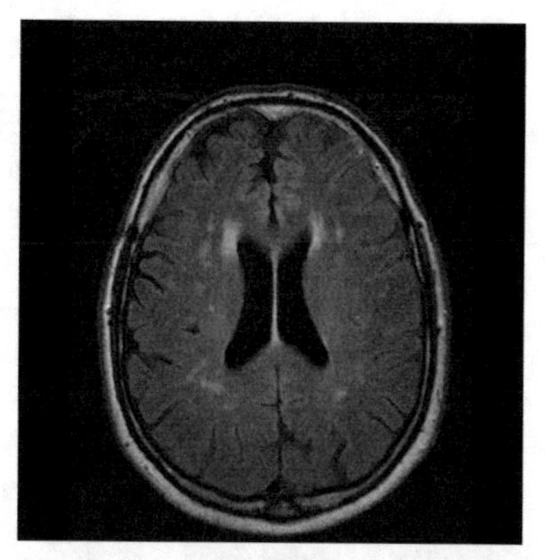

(a) Mild

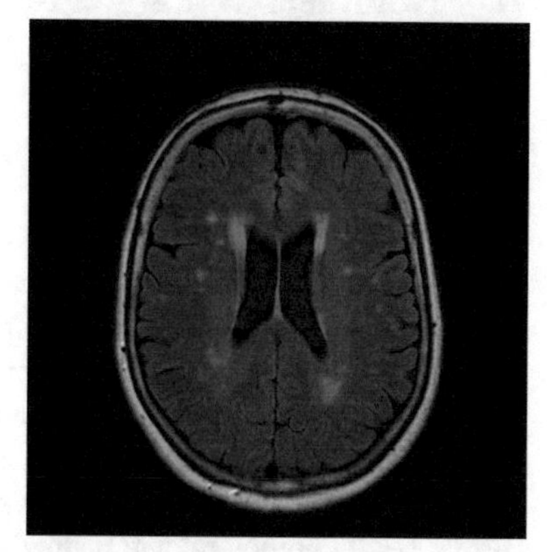

(b) Medium

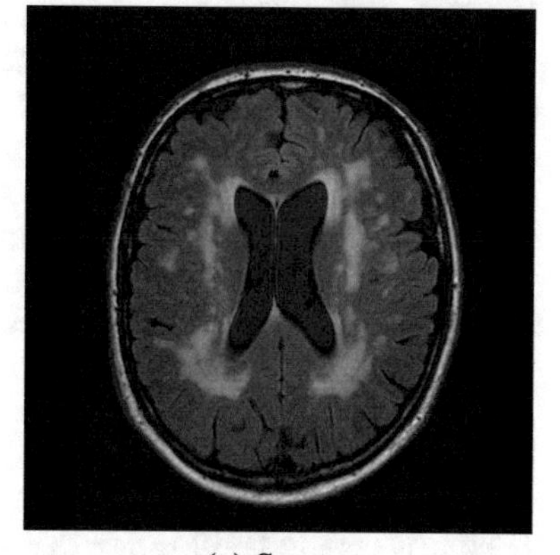

(c) Severe

Instead of using subjective rating as mild, medium, and severe or using integer scaling for the WMLs, which is subject to imprecision and difficulty in reproducing the assessment, the use of the 2D-FWRN can provide a measure of the WML patterns expressed on real numbers, which are precise and reproducible. The results consistently show the ability of the 2D-FWRN in scoring the WMLs in comparison with the reference ratings. The rationale for the effectiveness of the proposed method is that it can capture the nonlinear spatial distribution of the images in which the white matter hyperintensities exhibit a large influence over the entire space. The numerical values of the network properties can be useful for the precise WML quantification, potential features of spatial complex topology and shapes for MRI-based brain age prediction [126–129], and prediction of dementia [130, 131].

## 7.7 Nonlinear Dynamics of Photoplethysmogram in Parkinson's Disease

Photoplethysmography is an optical technique, which is known for its low cost and noninvasiveness. Photoplethysmogram (PPG) signals are measured by using a pulse oximeter which illuminates the skin and measures changes in light absorption [132]. A PPG sensor is applied on the skin surface to detect changes of the blood volume in peripheral circulation by capturing the changes of the intensity of infrared light traveling through biological tissues and absorbed by blood. PPG technology, including hardware and software, is still under ongoing development, and it has been used in many clinical physiological measurements and monitoring systems [133–135]. In general, a PPG is displayed as waveform data that have a pulsating alternating current (AC) component and a slowly changing direct current (DC) baseline [135]. The AC component reveals the blood volume variations being synchronized with the heartbeats, while the DC component reflects the influence of respiration, sympathetic nervous system activity as well as thermoregulation [133].

While most studies on the use of PPG focus on heart-rate events [136], fewer attempts have been reported on its application to neural systems. For the classification of Parkinson's disease and healthy control subjects, time series of stride intervals, which are signals of gait cycle durations recorded with multiple body sensors, are routinely used as a standard procedure for gaining insight into the neurodegenerative dynamics (https://www.physionet.org/physiobank/database/#gait).

Electrocardiography (ECG) and finger PPG were simultaneously used to evaluate the sympathetic vascular reactivity and the heart-rate response in patients with Parkinson's disease (PD) [137]. In fact, near-infrared light was reported to be of potential therapeutic effect in treating PD using cell cultures and animal models [138]. This study [139] shows that PPG signals are potentially useful for pattern analysis and differentiation between patients with PD and healthy control (HC) subjects with the use of nonlinear dynamics analysis methods and machine learning that are capable of capturing distinctive information of short-time signals to potentially overcome the limit of repeated long-time measurements.

The PPG dataset used for validating the nonlinear dynamics methods consisting of 112 HC subjects (young adults) and 32 PD patients was obtained from the Chaos Technology Research Laboratory, Shiga, Japan. The consent of the HC subjects and PD patients was obtained after an explanation of how the data would be used and that their personal information be kept confidential. The PPG signals were recorded for 2 min on an infrared sensor (UBIX Corporation) placed on the right earlobe of the subjects, who were at rest, at a sampling frequency of 200 Hz, low-pass filtered with a cutoff frequency of 30 Hz to remove noise from the external environment, and detrended. The total length of the PPG signals is about 36,000 time points. The first 2000 samples of the PPG signals were used as short-time data in this study. Figures 7.26 and 7.27 show examples of the PPG signals of the first 2000 samples and corresponding FRPs ($m = 1, c = 10$) of the HC subjects and PD patients, respectively. A noticeable difference between the PPG signals of the HC and PD is that the dicrotic notches are shown in the HC signals, known to be existing in young adults [140], but absent in the PD signals. The FRPs of the PD patients are sparser (show less amounts of recurrence) than those of the HC subjects.

To compute MSE, we set $m = 1$, $r = 0.5 \times \sigma$, where $\sigma$ is the standard deviation of the signal, and $\tau = 5$, 10, and 15. For the determination of TSME, $m = 1$, $r = 0.5 \times \sigma$, and $k_{\max} = 5$, 10, 15. For the construction of FRPs, $c = 5$, and 10, 20, and 20 features of the grayscale-level co-occurrence matrix [141] were extracted, which were similarly studied in the classification of gait dynamics between PD patients and HC subjects [66].

For the MSE, as shown in Table 7.23, the leave-one-out (LOO) cross-validations classified by the standard support vector machines (SVM) are all 87.88%, and by the least squares SVM (LS-SVM) [99] are 93.94, 100, and 100% using multiple scales, $\tau = 5$, 10, and 15, respectively. As results shown in Table 7.24 for the TSME, the LOO cross-validations classified by the support vector machines (SVM) are 86.36, 86.36, and 90.91% and by the LS-SVM are 96.97, 100, and 99.91% using multiple scales, $k = 5$, 10, and 15, respectively. For the FRPs, Table 7.25 shows the LOO cross-validations obtained from the SVM are 83.33 and 84.85%, and from the LS-SVM are all 100% for $c = 5$ and 10, respectively.

The LOO cross-validation, which was chosen in this experiment because of the small data size with particular reference to the PD cohort, results obtained from the LS-SVM using all the three types of features obtained from the three nonlinear dynamics analysis methods are consistently more favorable than those obtained from the standard SVM. In particular, the FRP gave 100% accuracy for either $c = 5$ or 10, where the selection of the number of fuzzy clusters in the FRP analysis is much less critical than those required by the other two nonlinear dynamic methods.

The average LS-SVM classification of short-time PPG features of the PD patients and HC subjects using various model parameters of the three nonlinear dynamics analysis methods is almost 100%, showing the potential real-time application of PPG on wearable devices for diagnosis and quantification of the disorder. It can be used to develop an automated system for monitoring tremors in patients with

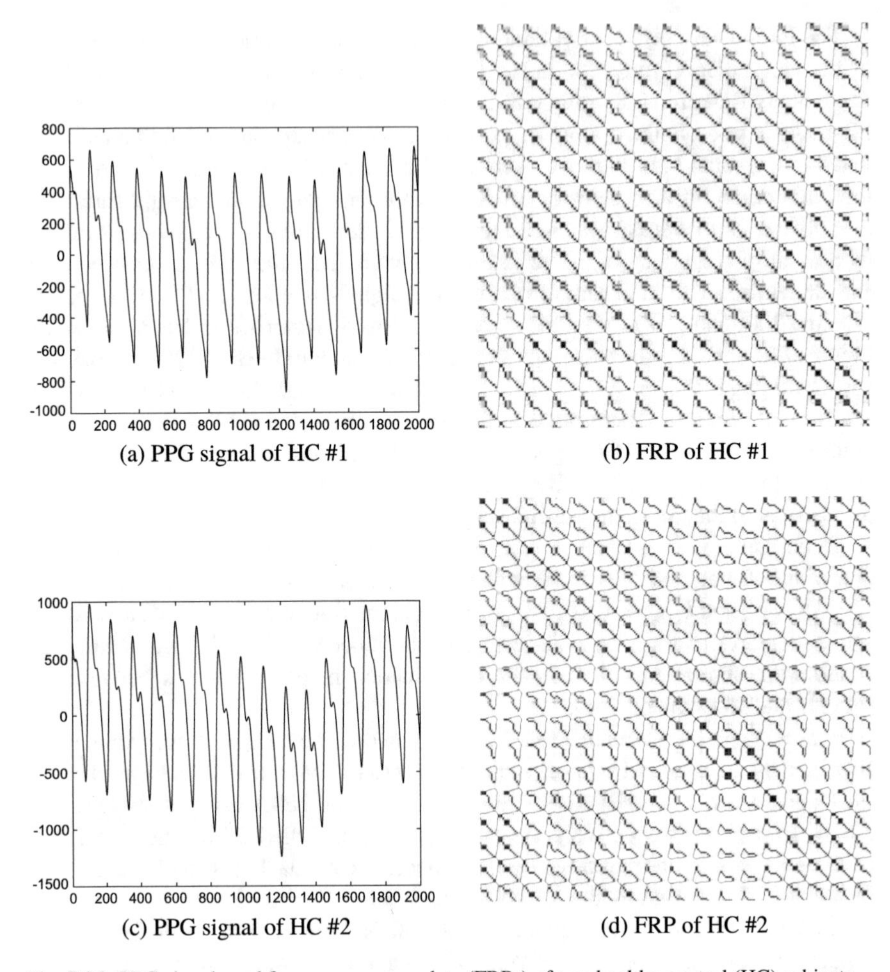


**Fig. 7.26** PPG signals and fuzzy recurrence plots (FRPs) of two healthy control (HC) subjects

PD and combined with other physiological signals obtained from multiple sensors for studying neurodegenerative diseases. If proving effective, the benefit of using PPG for studying disorders of the nervous system can be of many folds, such as providing a novel source of information for elucidating the underlying mechanism of the disorders, being of low cost, being easy to use, and relatively more convenient with the placement of multiple sensors on human subjects.

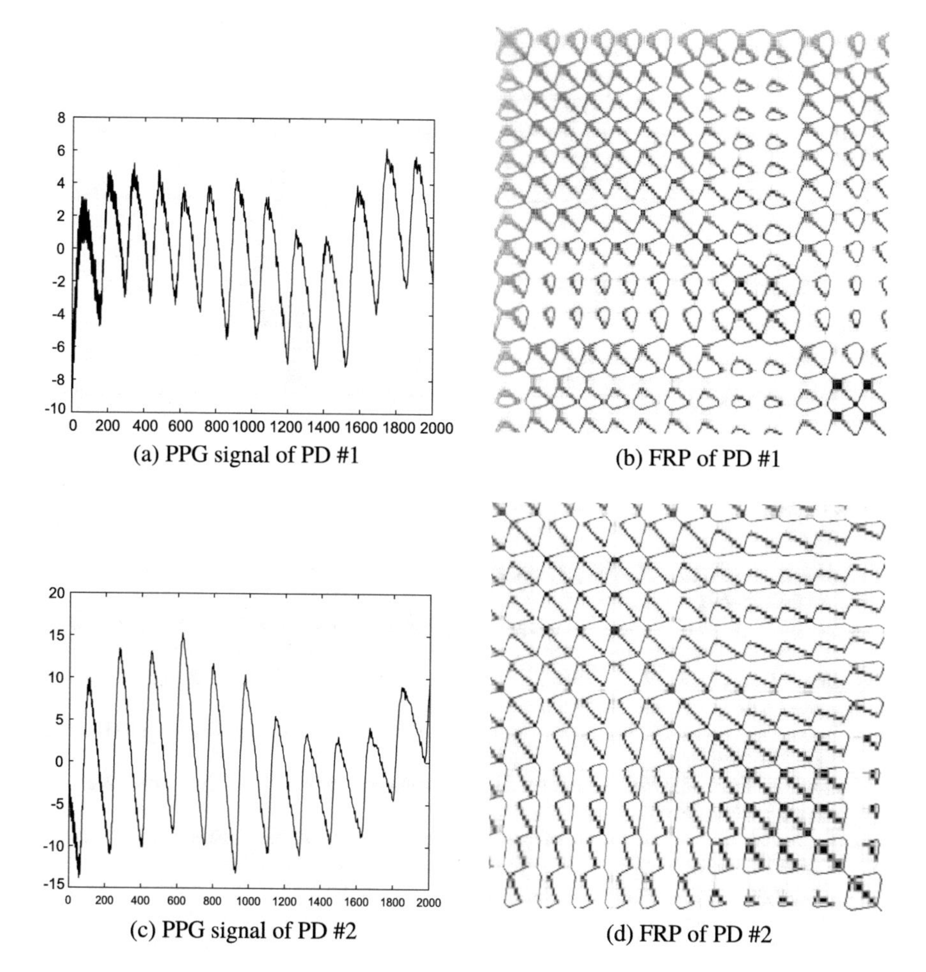


(a) PPG signal of PD #1

(b) FRP of PD #1

(c) PPG signal of PD #2

(d) FRP of PD #2

**Fig. 7.27** PPG signals and fuzzy recurrence plots (FRPs) of two Parkinson's disease (PD) patients

**Table 7.23** LOO-based accuracy (%) using MSE

| $\tau$ | SVM | LS-SVM |
|---|---|---|
| 5 | 87.88 | 93.94 |
| 10 | 87.88 | 100 |
| 15 | 87.88 | 100 |

**Table 7.24** LOO-based accuracy (%) using TSME

| $k$ | SVM | LS-SVM |
|---|---|---|
| 5 | 86.36 | 96.97 |
| 10 | 86.36 | 100 |
| 15 | 90.91 | 99.91 |

**Table 7.25** LOO-based accuracy (%) using FRP

| $c$ | SVM | LS-SVM |
|---|---|---|
| 5 | 83.33 | 100 |
| 10 | 84.85 | 100 |

# References

1. Arnold M et al (2017) Global patterns and trends in colorectal cancer incidence and mortality. Gut 66:683–691
2. Fan CW et al (2013) Cancer-initiating cells derived from human rectal adenocarcinoma tissues carry mesenchymal phenotypes and resist drug therapies. Cell Death Dis 4:e828
3. Sinicrope FA et al (2016) Molecular biomarkers in the personalized treatment of colorectal cancer. Clin Gastroenterol Hepatol 14:651–658
4. Zarkavelis G et al (2017) Current and future biomarkers in colorectal cancer. Ann Gastroenterol 30:613–621
5. Cuyle PJ, Prenen H (2017) Current and future biomarkers in the treatment of colorectal cancer. Acta Clin Belg 72:103–115
6. Rahman MR et al (2019) Identification of prognostic biomarker signatures and candidate drugs in colorectal cancer: Insights from aystems biology analysis. Medicina 55:E20
7. Chatterjee SB et al (2019) Lactosylceramide synthase $\beta$-1,4-GalT-V: A novel target for the diagnosis and therapy of human colorectal cancer. Biochem Biophys Res Commun 508:380
8. Letellier E et al (2017) Loss of Myosin Vb in colorectal cancer is a strong prognostic factor for disease recurrence. Br J Cancer 117:1689–1701
9. Lee PY et al (2018) Probing the colorectal cancer proteome for biomarkers: current status and perspectives. J Proteomics 118:93–105
10. Patel JN, Fong MK, Jagosky M (2019) Colorectal cancer biomarkers in the era of personalized medicine. J Pers Med 9:E3
11. Yaromina A, Krause M, Baumann M (2012) Individualization of cancer treatment from radiotherapy perspective. Mol Oncol 6:211–221
12. Sonke JJ, Belderbos J (2010) Adaptive radiotherapy for lung cancer. Semin Radiat Oncol 20:94e106
13. Logotheti S et al (2013) Functions, divergence and clinical value of TAp73 isoforms in cancer. Cancer Metastasis Rev 32:511–534
14. Lucena-Araujo AR et al (2015) High ΔNp73/TAp73 ratio is associated with poor prognosis in acute promyelocytic leukemia. Blood 126:2302–2306
15. Muller M et al (2005) TAp73/Delta Np73 influences apoptotic response, chemosensitivity and prognosis in hepatocellular carcinoma. Cell Death Differ 12:1564–1577
16. Di C et al (2013) Mechanisms, function and clinical applications of DNp73. Cell Cycle 15:1861–1867
17. Zhu W et al (2015) Expression and prognostic significance of TAp73 and ΔNp73 in FIGO stage I-II cervical squamous cell carcinoma. Oncol Lett 9:2090–2094

18. Uramoto H et al (2004) Expression of ΔNp73 predicts poor prognosis in lung cancer. Clin Cancer Res 10:6905–6911
19. Liu SS et al (2006) Expression of ΔNp73 and TAp73a independently associated with radiosensitivities and prognoses in cervical squamous cell carcinoma. Clin Cancer Res 12:3922–3927
20. Pfeifer D et al (2009) Protein expression following gamma-irradiation relevant to growth arrest and apoptosis in colon cancer cells. J Cancer Res Clin Oncol 135:1583–1592
21. Pham TD et al (2019) Image-based network analysis of DNp73 expression by immunohistochemistry in rectal cancer patients. Front Physiol 10:1551. https://doi.org/10.3389/fphys.2019.01551
22. Zhang W et al (2013) Network-based survival analysis reveals subnetwork signatures for predicting outcomes of ovarian cancer treatment. PLoS Comput Biol 9:e1002975
23. Parikh AP et al (2014) Network analysis of breast cancer progression and reversal using a tree-evolving network algorithm. PLoS Comput Biol 10:e1003713
24. Zhang F et al (2018) A novel heterogeneous network-based method for drug response prediction in cancer cell lines. Sci Rep 8:3355
25. Gladitz J, Klink B, Seifert M (2018) Network-based analysis of oligodendrogliomas predicts novel cancer gene candidates within the region of the 1p/19q co-deletion. Acta Neuropathol Commun 6:49
26. Ozturk K et al (2018) The emerging potential for network analysis to inform precision cancer medicine. J Mol Biol 430:2875–2899
27. Ruan J et al (2019) A novel algorithm for network-based prediction of cancer recurrence. Genomics 111:17–23
28. Trial SRC et al (1997) Improved survival with preoperative radiotherapy in resectable rectal cancer. N Engl J Med 8:980–987
29. London S (2017) Common causes of death predominate among long-term colorectal cancer survivors. ASCO Post 25
30. Walther A et al (2009) Genetic prognostic and predictive markers in colorectal cancer. Nat Rev Cancer 9:489–499
31. Voon PJ, Kong HL (2011) Tumour genetics and genomics to personalise cancer treatment. Ann Acad Med Singap 40:362–368
32. Amelio I et al (2014) TAp73 promotes anabolism. Oncotarget 5:12820–12934
33. Stantic M et al (2015) TAp73 suppresses tumor angiogenesis through repression of proangiogenic cytokines and HIF-1$\alpha$ activity. Proc Natl Acad Sci USA 112:220–225
34. Depression (2018) National institute of mental health. https://www.nimh.nih.gov/health/topics/depression/index.shtml, Accessed 16 June 2019
35. Depression (2019) Mayo clinic. https://www.mayoclinic.org/diseases-conditions/depression/symptoms-causes/syc-20356007, Accessed 16 June 2019
36. Ferrari AJ et al (2010) Burden of depressive disorders by country, sex, age, and year: findings from the global burden of disease study 2010. PLoS Med 10:10e1001547
37. Bhandari S (2017) Untreated depression: WebMD medical reference. https://www.webmd.com/depression/guide/untreated-depression-effects#1, Accessed 16 June 2019
38. Briggs R et al (2018) What is the prevalence of untreated depression and death ideation in older people? data from the Irish longitudinal study on aging. Int Psychogeriatr 30:1393–1401
39. Wichers M et al (2016) Critical slowing down as a personalized early warning signal for depression. Psychother Psychosom 85:114–116
40. Scheffer M et al (2009) Early-warning signals for critical transitions. Nature 461:53–59
41. Van de Leemput IA et al (2014) Critical slowing down as early warning for the onset and termination of depression. Proc Natl Acad Sci USA 111:87–92
42. Beard C et al (2016) Network analysis of depression and anxiety symptom relationships in a psychiatric sample. Psychol Med 46:3359–3369
43. McElroy E et al (2019) Structure and connectivity of depressive symptom networks corresponding to early treatment response. EClinicalMedicine 8:29–36
44. Wichers M et al (2017) Mental disorders as networks: some cautionary reflections on a promising approach. Soc Psychiatry Psychiatr Epidemiol 52:143–145

45. Schweren L et al (2018) Assessment of symptom network density as a prognostic marker of treatment response in adolescent depression. JAMA Psychiatry 75:98–100
46. Groot PC (2010) Patients can diagnose too: how continuous self-assessment aids diagnosis of, and recovery from, depression. J Ment Heal 19:352–362
47. Kossakowski JJ et al (2017) Data from 'critical slowing down as a personalized early warning signal for depression'. J Open Psychol Data 5:1
48. Cichocki A et al (2015) Tensor decompositions for signal processing applications: from two-way to multiway component analysis. IEEE Signal Process Mag 32:145–163
49. Harshman RA (1970) Foundations of the PARAFAC procedure: models and conditions for an "explanatory" multi-modal factor analysis. UCLA Work Pap Phon 16:1–84
50. Bro R (1997) PARAFAC. Tutorial and applications. Chemom Intell Lab Syst 38:149–171
51. Pham TD, Yan H (2018) Tensor decomposition of gait dynamics in Parkinson's disease. IEEE Trans Biomed Eng 65:1820–1827
52. Andersson CA, Bro R (2000) The N-way toolbox for MATLAB. Chemom Intell Lab Syst 52:1–4
53. Chinta SJ, Andersen JK (2005) Dopaminergic neurons. Int J Biochem Cell Biol 37:942–946
54. Statistics on Parkinson's (2017) Parkinson's disease foundation. http://www.pdf.org/parkinson_statistics
55. Rizek P, Kumar N, Jog MS (2016) An update on the diagnosis and treatment of Parkinson disease. CMAJ 188:1157–1165
56. Elkouzi A (2019) What is parkinson's? Parkinson's Foundation. https://parkinson.org/understanding-parkinsons/what-is-parkinsons
57. DeMaagd G, Philip A (2015) Parkinson's disease and its management: part 1: disease entity, risk factors, pathophysiology, clinical presentation, and diagnosis. P&T 40:504–532
58. Hariz M, Blomstedt P, Zrinzo L (2013) Future of brain stimulation: new targets, new indications, new technology. Mov Disord 28:1784–1792
59. Hariz M (2014) Deep brain stimulation: new techniques. Park Relat Disord 20(Suppl 1):S192–S196
60. Emamzadeh FN, Surguchov A (2018) Parkinson's disease: biomarkers, treatment, and risk factors. Front Neurosci 12:612
61. Oertel WH (2017) Recent advances in treating Parkinson's disease. F1000Research 6:260
62. Hausdorff JM (2009) Gait dynamics in Parkinson's disease: common and distinct behavior among stride length, gait variability, and fractal-like scaling. Chaos 19:026113
63. Chen PH et al (2013) Gait disorders in Parkinson's disease: assessment and management. Int J Gerontol 7:189–193
64. Zeng W, Wang C (2016) Parkinson's disease classification using gait analysis via deterministic learning. Neurosci Lett 633:268–278
65. Ren P et al (2017) Gait rhythm fluctuation analysis for neurodegenerative diseases by empirical mode decomposition. IEEE Trans Biomed Eng 64:52–60
66. Pham TD (2018) Texture classification and visualization of time series of gait dynamics in patients with neuro-degenerative diseases. IEEE Trans Neural Syst Rehabil Eng 26:188–196
67. Kribus-Shmiel L et al (2018) How many strides are required for a reliable estimation of temporal gait parameters? Implementation of a new algorithm on the phase coordination index. PLoS ONE 13:e0192049
68. Giancardo L et al (2016) Computer keyboard interaction as an indicator of early Parkinson's disease. Sci Rep 6:34468
69. Tavares TAL et al (2005) Quantitative measurements of alternating finger tapping in Parkinson's disease correlate with UPDRS motor disability and reveal the improvement in fine motor control from medication and deep brain stimulation. Mov Disord 20:1286–1298
70. Pham TD (2016) Fuzzy recurrence plots. EPL 116:50008
71. Pham TD et al (2019) Classification of short time series in early Parkinson's disease with deep learning of fuzzy recurrence plots. IEEE/CAA J Autom Sin 6:1306–1317
72. Che Z et al (2018) Recurrent neural networks for multivariate time series with missing values. Sci Rep 8:6085

73. Hinton G et al (2012) Deep neural networks for acoustic modeling in speech recognition: the shared views of four research groups. IEEE Signal Process Mag 29:82–97
74. Sutskever I, Vinyals O, Le QV (2014) Sequence to sequence learning with neural networks. In: Ghahramani Z et al (eds) Advances in neural information processing systems. Curran Associates, New York, pp 3104–3112
75. Bahdanau D, Cho K, Bengio Y (2015) Neural machine translation by jointly learning to align and translate. In: Bengio Y, LeCun Y (eds) Proceedings of 3rd international conference on learning representations
76. Malhotra P et al (2017) TimeNet: pre-trained deep recurrent neural network for time series classification. In: 25th European symposium on artificial neural networks, Computational intelligence and machine learning, pp 607–612
77. Mehdiyeva N et al (2017) Time series classification using deep learning for process planning: a case from the process industry. Procedia Comput Sci 114:242–249
78. Shi X et al (2015) Convolutional LSTM network: a machine learning approach for precipitation nowcasting. Adv Neural Inf Process Syst 28:802–810
79. Cui Z, Chen W, Chen Y (2016) Multi-scale convolutional neural networks for time series classification. In: CoRR, arXiv:1603.06995
80. Wang Z, Yan W, Oates T (2017) Time series classification from scratch with deep neural networks: a strong baseline. In: 2017 International joint conference on neural networks, pp 1578–1585
81. Karim F et al (2018) LSTM fully convolutional networks for time series classification. IEEE Access 6:1662–1669
82. Fawaz HI et al (2019) Deep learning for time series classification: a review. In: Data mining and knowledge discovery, pp 1–47
83. Hochreiter S, Schmidhuber J (1997) Long short-term memory. Neural Comput 9:1735–1780
84. Graves A, Schmidhuber J (2005) Framewise phoneme classification with bidirectional LSTM and other neural network architectures. Neural Netw 18:602–610
85. Graves A, Jaitly N, Mohamed A (2013) Hybrid speech recognition with deep bidirectional LSTM. 2013 IEEE workshop on automatic speech recognition and understanding. Czech Republic, Olomouc, pp 273–278
86. Zazo R et al (2016) Language Identification in short utterances using long short-term memory (LSTM) recurrent neural networks. PLOS ONE 11:e0146917
87. Zhenyang L et al (2018) VideoLSTM convolves, attends and flows for action recognition. Comput Vis Image Underst 166:41–50
88. Mikolov T et al (2011) Extensions of recurrent neural network language model. 2011 IEEE international conference on acoustics. Speech and Signal Processing, Prague, Czech Republic, pp 5528–5531
89. Pascanu R, Mikolov T, Bengio Y (2013) On the difficulty of training recurrent neural networks. In: Proceedings of the 30th international conference on international conference on machine learning, Atlanta, GA, USA, pp III-1310–III-1318
90. Greff K et al (2017) LSTM: a search space odyssey. IEEE Trans Neural Netw Learn Syst 28:2222–2232
91. NeuroQWERTY MIT-CSXPD Dataset. PhysioNet. https://www.physionet.org/physiobank/database/nqmitcsxpd/
92. Martinez-Martin P et al (1994) Unified Parkinson's disease rating scale characteristics and structure. Coop Multicentric Group Mov Disord 9:76–83
93. Pham TD (2018) Pattern analysis of computer keystroke time series in healthy control and early-stage Parkinson's disease subjects using fuzzy recurrence and scalable network features. J Neurosci Methods 307:194–202
94. Kantz H, Schreiber T (2004) Nonlinear time series analysis. University Press, Cambridge
95. Szegedy C et al (2015) Going deeper with convolutions. In: 2015 IEEE conference on computer vision and pattern recognition, Boston MA, pp 1–9
96. Krizhevsky A, Sutskever I, Hinton GE (2017) ImageNet classification with deep convolutional neural networks. Commun ACM 60:84–90

97. MathWork, Classify time series using wavelet analysis and deep learning. https://mathworks.com/help/deeplearning/examples/signal-classification-with-wavelet-analysis-and-convolutional-neural-networks.html#d117e22181. Accessed 25 April 2019
98. Schuster M, Paliwal KK (1997) Bidirectional recurrent neural networks. IEEE Trans Signal Process 45:2673–2681
99. Suykens JAK et al (2002) Least squares support vector machines. World Scientific, Singapore
100. Shimoyama T, Ninchoji T, Uemura K (1990) The finger-tapping test. A Quant Anal Arch Neurol 47:681–684
101. Marwan N et al (2007) Recurrence plots for the analysis of complex systems. Phys Rep 438:237–329
102. Fabretti A, Ausloos M (2005) Recurrence plot and recurrence quantification analysis techniques for detecting a critical regime. Int J Mod Phys C 16:671–706
103. Fraser A, Swinney H (1986) Independent coordinates for strange attractors from mutual information. Phys Rev A 33:1134–1140
104. Kennel M, Brown R, Abarbanel H (1992) Determining embedding dimension for phase space reconstruction using a geometrical construction. Phys Rev A 45:3403–3411
105. Iwanski JS, Bradley E (1998) Recurrence plots of experimental data: to embed or not to embed? Chaos 8:861
106. Calvano SE et al (2005) A network-based analysis of systemic inflammation in humans. Nature 437:1032–1037
107. Hopkins AL (2008) Network pharmacology: the next paradigm in drug discovery. Nat Chem Biol 4:682–690
108. Barabasi AL, Gulbahce N, Loscalzo J (2011) Network medicine: a network-based approach to human disease. Nat Rev Genet 12:56–68
109. Chan SY, Loscalzo J (2012) The emerging paradigm of network medicine in the study of human disease. Circ Res 111:359–374
110. Ideker T, Nussinov R (2017) Network approaches and applications in biology. PLoS Comput Biol 13:e1005771
111. Mhyre TR et al (2012) Parkinson's disease. Sub-Cell Biochem 65:389–455
112. Lowenstein DH, Martin JB, Hauser SL (2012) Approach to the patient with neurologic disease. In: Longo DL et al (eds) Harrison's principles of internal medicine, Chapter 367, 18th edn. McGraw-Hill, New York
113. Hausdorff JM et al (1997) Altered fractal dynamics of gait: reduced stride-interval correlations with aging and Huntington's disease. J Appl Physiol 82:262–269
114. Hausdorff JM et al (2000) Dynamic markers of altered gait rhythm in amyotrophic lateral sclerosis. J Appl Physiol 88:2045–2053
115. Wu J, Krishnan S (2010) Statistical analysis of gait rhythm in patients with Parkinson's disease. IEEE Trans Neural Syst Rehabil Eng 18:150–158
116. Khorasani A, Daliri MR (2014) HMM for classification of Parkinson's disease based on the raw gait data. J Med Syst 38:147
117. Zeng W, Wang C (2015) Classification of neurodegenerative diseases using gait dynamics via deterministic learning. Inf Sci 317:246–258
118. Hoehn M, Yahr M (1967) Parkinsonism: onset, progression and mortality. Neurology 17:427–442
119. Unser M (1986) Sum and difference histograms for texture classification. IEEE Trans Pattern Anal Mach Intell 8:118–125
120. Yang X (2012) Ultrasound GLCM texture analysis of radiation-induced parotid-gland injury in head-and-neck cancer radiotherapy: an in vivo study of late toxicity. Med Phys 39:5732–5739
121. Kim KW, MacFall JR, Payne ME (2008) Classification of white matter lesions on magnetic resonance imaging in elderly persons. Biol Psychiatry 64:273–280
122. Mantyla R et al (1997) Variable agreement between visual rating scales for white matter hyperintensities on MRI: comparison of 13 rating scales in a poststroke cohort. Stroke 28:1614–1623

123. Melhem ER et al (2003) Defining thresholds for changes in size of simulated T2-hyperintense brain lesions on the basis of qualitative comparisons. AJR Am J Roentgenol 180:65–69
124. Pham TD (2019) Quantification of white matter lesions on brain MRI with 2D fuzzy weighted recurrence networks. 9th international IEEE/EMBS conference on neural engineering, 20–23 March 2019. CA, USA, San Francisco, pp 110–113
125. Pham TD et al (2011) The hidden-Markov brain: comparison and inference of white matter hyperintensities on magnetic resonance imaging (MRI). J. Neural Eng 8:016004
126. Wang B, Pham TD (2011) MRI-based age prediction using hidden Markov models. J Neurosci Methods 199:140–145
127. Lin L et al (2016) Predicting healthy older adult's brain age based on structural connectivity networks using artificial neural networks. Comput Methods Programs Biomed 125:8–17
128. Pham TD, Abe T, Oka Y, Chen YF (2015) Measures of morphological complexity of gray matter on magnetic resonance imaging for control age grouping. Entropy 17:8130–8151
129. Franke K et al (2018) Premature brain aging in humans exposed to maternal nutrient restriction during early gestation. NeuroImage 173:460–471
130. Chen Y, Pham TD (2013) Development of a brain MRI-based hidden Markov model for dementia recognition. Biomed Eng Online 12:S2
131. Ramaniharan AK, Manoharan SC, Swaminathan R (2016) Laplace Beltrami eigen value based classification of normal and Alzheimer MR images using parametric and non-parametric classifiers. Expert Syst Appl 59:208–216
132. Shelley K, Shelley S (2001) Pulse oximeter waveform: photoelectric plethysmography. In: Lake C, Hines R, Blitt C (eds) Clinical monitoring. W.B. Saunders, Philadelphia, pp 420–428
133. Allen J (2007) Photoplethysmography and its application in clinical physiological measurement. Physiol Meas 28:R1–R39
134. Bartels K, Thiele RH (2015) Advances in photoplethysmography: beyond arterial oxygen saturation. Can J Anaesth 62:1313–1328
135. Fallet S, Vesin JM (2017) Robust heart rate estimation using wrist-type photoplethysmographic signals during physical exercise: an approach based on adaptive filtering. Physiol Meas 38:155
136. Rajaguru H, Prabhakar SK (2015) A comprehensive review on photoplethysmography and its application for heart rate turbulence clinical diagnosis. Adv Sci Lett 21:3602–3604
137. Roy S, Srivastava AK, Jaryal AK, Deepak KK (2015) Cardiovascular responses during cold pressor test are different in Parkinson disease and multiple system atrophy with parkinsonism. Clin Auton Res 25:219–224
138. Quirk BJ et al (2012) Therapeutic effect of near infrared (NIR) light on Parkinson's disease models. Front Biosci 4:818–823
139. Pham TD, Oyama-Higa M (2018) Nonlinear dynamics analysis of short-time photoplethysmogram in Parkinson's disease. In: 2018 IEEE international conference on fuzzy systems, 08–13 July 2018, Rio de Janeiro, Brazil, pp 1749–1754
140. Shi P et al (2009) Insight into the dicrotic notch in photoplethysmographic pulses from the finger tip of young adults. J Med Eng Technol 33:628–633
141. Haralick RM, Shanmugam K, Dinstein I (1973) Textural features of image classification. IEEE Trans Syst, Man Cybern 3:610–621

# Index




T. D. Pham, *Fuzzy Recurrence Plots and Networks with Applications in Biomedicine*, https://doi.org/10.1007/978-3-030-37530-0

**P**



**R**



**S**



**T**



**U**



**V**



**W**

Printed in the United States
By Bookmasters